Must Obey	Don't Do It	Risk of Danger	Safety Way
Mandatory	Prohibition	Caution	Safe condition

Safety signs reproduced with permission of Jencons [Scientific] Limited, Leighton Buzzard, UK.

Chemical safety matters

Editorial Group

PETER H. BACH,
Faculty of Science,
Polytechnic of East London,
London, E15 4LZ
U.K.

JOHN A. HAINES,
International Programme on
Chemical Safety,
1211 Geneva 27,
Switzerland

BLAINE C. McKUSICK,
formerly of:
E.I. Du Pont de Nemours & Co. Inc.,
Wilmington, Delaware, DE 19898,
U.S.A

STANLEY S. BROWN,
Regional Laboratory for Toxicology,
Birmingham, B18 7QH
U.K.

ROBERT M. JOYCE,
formerly of:
E.I. Du Pont de Nemours & Co. Inc.,
Wilmington, Delaware, DE 19898,
U.S.A.

HOWARD G. J. WORTH
King's Mill Hospital,
Sutton-in-Ashfield,
Nottinghamshire, NG17 4JL
U.K.

Chemical Safety Matters

IUPAC – IPCS
INTERNATIONAL UNION OF PURE AND APPLIED CHEMISTRY
INTERNATIONAL PROGRAMME ON CHEMICAL SAFETY

International Union of Pure and Applied Chemistry

United Nations Environment Programme

International Labour Organisation

World Health Organization

Published by Cambridge University Press on behalf of the World Health Organization and the International Union of Pure and Applied Chemistry.

CAMBRIDGE UNIVERSITY PRESS

Published by the Press Syndicate of the University of Cambridge
The Pitt Building, Trumpington Street, Cambridge CB2 1RP
40 West 20th Street, New York, NY 10011-4211, USA
10 Stamford Road, Oakleigh, Victoria 3166, Australia

© the World Health Organization and the International Union of
Pure and Applied Chemistry 1992

First published 1992

Printed in Great Britain at the University Press, Cambridge

A catalogue record for this book is available from the British Library

Library of Congress cataloguing in publication data
Chemical Safety Matters
WHO/IUPAC
    p.   cm.
Includes bibliographical references and index.
ISBN 0 521 41375 3
1. Chemical laboratories – Safety measures.   2. Chemical
laboratories – Waste disposal.
QD63.5.C47   1992
542′.028′9 – dc20   91-28312   CIP

ISBN 0 521 41375 3 paperback

This work contains the collective views of an international group of experts and does not necessarily represent the decisions or stated policy of the International Union of Pure and Applied Chemistry, the United Nations Environment Programme, the International Labour Organisation, the World Health Organization or Cambridge University Press. It is provided as advice, has no legally binding status and is not intended either to be used in the regulatory process in any specific country or to imply an endorsement of any commercial product.
    Every reasonable effort has been made to ensure that the text of this work is free from error. None the less no warranty is made that errors do not occur, and the editors and publisher disclaim all liability for direct or consequential damages resulting from the use of this work.

UP

Contents

List of tables		ix
Preface		xi
Units, Nomenclature and Abbreviations		xvii
	Introduction	1
PART I	**Safe Working Procedures and Protective Equipment**	**11**
1.	General recommendations for safe practices in laboratories	13
2.	Protective apparel, safety equipment, emergency procedures, and first aid	21
3.	Laboratory ventilation	50
4.	Design requirements and use of electrically-powered laboratory apparatus	63
PART II	**Safe Storage and Use of Hazardous Chemicals**	**75**
5.	Procedures for the procurement, storage and distribution of chemicals	77
6.	Procedures for working with substances that pose hazards because of acute toxicity, chronic toxicity, or corrosivity	88
7.	Procedures for working with substances that pose flammable or explosive hazards	111
8.	Procedures for working with gases at pressures above or below atmospheric	131
PART III	**Safe Storage and Disposal of Waste Chemicals**	**155**
9.	Recovery, recycling, and reuse of laboratory chemicals	157
10.	Procedures for laboratory destruction of hazardous chemicals	162
11.	Disposal of explosives from laboratories	198
12.	A waste management system for laboratories	201
13.	Identification, classification and segregation of laboratory waste	206
14.	Storage of laboratory waste	213
15.	Disposal of chemicals in the sanitary sewer system	216
16.	Incineration of hazardous chemicals	220
17.	Transportation of hazardous chemicals	227
18.	Disposal of hazardous chemicals in landfills	232
19.	Disposal of chemically contaminated waste from life-science laboratories	236

Appendix A	Water-reactive chemicals	241
Appendix B	Peroxide-forming chemicals	242
Appendix C	Pyrophoric chemicals	245
Appendix D	Potentially explosive chemicals combinations	246
Appendix E	Incompatible chemicals	248
Appendix F	pH ranges for precipitating hydroxides of cations	251
Appendix G	Guidelines for disposal of chemicals in the sanitary sewer	252
	Glossary	254
	Bibliography	270
	Chemical index	273
	General index	279

Tables

Table 2.1	Resistance of common glove materials to chemicals	24
Table 7.1	Flash points, boiling points, ignition temperatures, and flammable limits of some common laboratory chemicals	114
Table 10.1	Relative toxicities of cations	179
Table 10.2	Hazardous properties of anions	180
Table 10.3	Precipitation of sulfides	182
Table 16.1	Calculated temperatures (°C) for destruction of selected compounds with efficiencies of 99% and 99.99% for residence times of 1 or 2 seconds	224
Table 18.1	Examples of non-hazardous laboratory wastes	234
Table B.1	Types of chemicals that may form peroxides	243
Table B.2	Common peroxide-forming chemicals	244
Table D.1	Shock-sensitive compounds	246
Table D.2	Potentially explosive combinations of some common reagents	247
Table E.1	General classes of incompatible chemicals	248
Table E.2	Specific chemical incompatibilities	249
Table F.1	pH ranges for precipitation of metal hydroxides and oxides	251
Table G.1	Non-hazardous organic and inorganic chemicals suitable for sanitary sewer disposal	252

Preface

In 1975, a planning group of the *United States National Academy of Sciences* concluded that there was need to recommend procedures for the safe use and disposal of harmful substances in chemical laboratories. The intention was to encompass all types of laboratories where chemicals are used, and to include not only hazards from acute or chronic chemical toxicity (including carcinogenicity), but also general hazards such as fires and explosions. However, special hazards from physical sources, such as radioactivity or lasers, or from biological sources, such as pathogenic bacteria or viruses, would not be considered in detail. The recommendations were to be limited to laboratory operations, where relatively small quantities of chemicals are used; they were not intended to apply to larger-scale operations such as pilot plant or manufacturing processes.

With this remit, a committee chaired by Herbert O. House was convened, comprising representatives of academic, governmental, and industrial laboratories. The first phase of the study was a survey of the safety practices which were being followed by representative laboratories. The combined experience and knowledge of individuals from those laboratories served as the basis for the procedures which were recommended in a study report published in 1981 by the *National Academy Press* under the title *Prudent Practices for Handling Hazardous Chemicals in Laboratories*.

As that report was being completed, elaborate regulations on the management of hazardous wastes, prepared by the *United States Environmental Protection Agency*, became effective. Many laboratories - academic, governmental, and industrial - found difficulty in understanding and complying with these regulations.

Accordingly, a committee chaired by Robert A. Alberty, also under the aegis of the National Academy of Sciences, undertook a study of the disposal of chemicals from laboratories. The objective was to develop guidelines and procedures for such disposal that would be safe and environmentally acceptable. As with the first report, the procedures were framed for laboratory operations, in which many types of chemicals are used in relatively small quantities; they were not intended to be applicable to manufacturing operations, where, in general, a smaller variety of chemicals is used on a much larger scale. The recommendations of the committee were published by the National Academy Press in 1983 under the title *Prudent Practices for the Disposal of Chemicals from Laboratories*. Concern about the proper use of chemicals in laboratories and their disposal was - and is - worldwide. In 1984, the President of the *International Union of Pure and Applied Chemistry* (IUPAC), William G. Schneider, suggested that the key information on these problems could be made more readily available to many countries by consolidating the two National Academy reports and omitting material that was specific to the United States of America. The abridgement involved deleting the extensive discussions of United States regulations, but indicating in general terms the kinds of practices and regulations that are common in other countries. In addition, parts of the two reports have been rearranged, updated, and some new procedures added so as to enhance the usefulness of the book both in teaching and in laboratory practice.

The *International Programme on Chemical Safety* (IPCS) - a joint venture of the *International Labour Office* (ILO), the *United Nations Environment Programme* (UNEP), and the *World Health Organization* (WHO) - was established in 1980 as the scientific response of the United Nations system to the challenges presented to the health of present and future generations, and to the quality of the environment, by the widespread utilisation of chemicals throughout the world. The objectives of the IPCS are two-fold: firstly, to catalyse and coordinate activities in relationship to chemical safety, particularly as far as the assessment of the risks to human health and to the environment from exposure to chemicals is concerned; secondly, to strengthen capabilities so as to ensure the protection of human health and of the environment from the deleterious effects of chemicals in all aspects of their production, transportation, handling, use, and disposal. In fulfilling its objectives, the IPCS enlists the support of the technical community, part of which is represented through IUPAC, an organization having official non-government organization status with the WHO. Policy guidance for the work of IPCS is provided through a Programme Advisory Committee; at its fifth meeting, in October 1986, when it was reviewing work initiated earlier on quality assurance in laboratory studies, this committee emphasised the need to provide guidance on protection of laboratory workers' health and their safety, and on safe disposal of their laboratory toxic wastes. Attention was drawn to the need to provide a practical manual for use of laboratories in developing countries.

During consultations between the President of the Clinical Chemistry Division of IUPAC and the Manager of the IPCS, it became evident that the efforts of IUPAC to consolidate the experience from North America on the laboratory handling and disposal of hazardous chemicals could form the basis of such a practical manual and provide guidance to all countries. Moreover, through IPCS, it would be possible to obtain inputs from parts of the world where IUPAC is not represented. In the context of an IPCS meeting on Recommended Operating Procedures for Laboratories, held in Finland, October 1987, the draft IUPAC document on prudent practices for handling and disposal of hazardous laboratory chemicals was examined with a view to how it could be adapted to meet the needs of IPCS for a practical manual and for global guidance. Comments from this meeting were incorporated into the draft document. The IPCS mechanisms for peer review were then utilised in order to obtain a broad consensus and to ensure that the document took fully into consideration experience and conditions in all regions of the world. IUPAC and the IPCS designated Howard G.J. Worth and Peter Bach respectively to assist in editing the document. A joint IPCS/IUPAC Editorial Group met in the United Kingdom, in December 1988, in order to review comments and input from referees and to prepare the text for publication. While this publication is based in part on the two National Academy of Sciences' publications, quoted above, it does not contain the United States of America's regulations and background, to which the reader is referred and which are available from the National Academy Press, Washington D.C.

The names of those who were members of one or both of the National Academy committees are listed below, together with those who were invited to serve as referees of the draft abridgement at various stages. Many others contributed comments, advice, help with illustrations and encouragement, notably William Spindel, a member of the National Research Council staff and the Director of both studies, and his successor, Robert M. Simon; Maurice Williams, Executive Secretary of the International Union of Pure and Applied Chemistry; and Michel Mercier, Manager of the International Programme on Chemical Safety. We extend our warm appreciation to them all.

Peter H. Bach, London, U.K.
Stanley S. Brown, Birmingham, West Midlands, U.K.
John A. Haines, Geneva, Switzerland
Robert M. Joyce, Sun City Center, Florida, U.S.A.
Blaine C. McKusick, Wilmington, Delaware, U.S.A.
Howard G.J. Worth, Sutton-in-Ashfield, Nottinghamshire, U.K.

Members of the U.S. National Academy of Sciences Committee on Hazardous Substances in the Laboratory

Robert A. Alberty, Massachusetts Institute of Technology
Edwin D. Becker, National Institutes of Health
Jerome A. Berson, Yale University
Elkan Blout, Harvard University Medical School
Larry I. Bone, Dow Chemical USA
Theodore L. Cairns, Pennsylvania
William G. Bauben, University of California, Berkeley
Robert W. Day, University of Washington
Alain DeCleve, Stanford University
Thomas S. Ely, Eastman Kodak Company
Ronald W. Estabrook, University of Texas Health Science Center
Margaret C. Etter, 3M Company
John M. Fresina, Massachusetts Institute of Technology
Gerhart Friedlander, Brookhaven National Laboratory
Irving H. Goldberg, Harvard Medical School
Leon Gordis, Johns Hopkins School of Public Health
Anna J. Harrison, Mount Holyoke College
Clayton Hathaway, Monsanto Company
Clayton H. Heathcock, University of California, Berkeley
Herbert O. House, Georgia Institute of Technology
Donald M. Jerina, National Institute of Arthritis, Metabolism & Digestive Diseases
Robert M. Joyce, Hockessin, Delaware
Michael M. King, George Washington University
Irvin L. Klundt, Aldrich Chemical Company
Marvin Kuschner, State University of New York Health Science Center
Ralph R. Langner, Dow Chemical Company
Blaine C. McKusick, E.I. Du Pont de Nemours & Company Inc.
John F. Meister, Southern Illinois University
William G. Mikell, E.I. Du Pont de Nemours & Company Inc.
Elizabeth C. Miller, University of Wisconsin
Robert A. Neal, Vanderbilt University
J.E. Rall, National Institute of Arthritis, Metabolism & Digestive Diseases
George Roush Jr., Monsanto Company
Adel F. Sarofim, Massachusetts Institute of Technology
William P. Schaefer, California Institute of Technology
Alfred W. Shaw, Shell Development Company
Howard E. Simmons, E.I. Du Pont de Nemours & Company Inc.
Ralph G. Smith, University of Michigan
Fay M. Thompson, University of Minnesota
Max Tishler, Wesleyan University
P. Christian Vogel, BASF Wyandotte Corporation
Kenneth L. Williamson, Mount Holyoke College

Invited Referees

A. Akintonwa, University of Lagos, Lagos, Nigeria
Margarida de Barros, Organisation for Economic Development and Cooperation, Paris, France
E.L. Brisson, US Food and Drug Administration
M. Castegnaro, International Agency for Research on Cancer, Lyon, France
Bahira Fahim, Poison Control Centre, Cairo, Egypt
George Frost, Robens Institute of Health & Safety, Guildford, United Kingdom
Marika Geldmacher-von Mallinckrodt, Senatskommission der Deutschen Forschungsgemeinschaft für Klinisch-Toxikilogische Analytik, Erlangen, Germany
C. A. Gotelli, Centro de Investigaciaones Toxicologicas, Buenos Aires, Argentina
A.N.P. van Heijst, The Netherlands
S. Hernberg, Institute of Occupational Health, Helsinki, Finland
J.R. Hickman, Environmental Health Directorate, Health & Welfare, Ottawa, Canada
Filiz Hincel, Hacettepe University, Ankara, Turkey
F. Jardine, Polytechnic of East London, London
J.O. Järvisalo, Social Insurance Institution, Turku, Finland
R. Kroes, National Institute of Public Health & Environmental Protection, Bilthoven, The Netherlands
Suvi Lehtinen, Institute of Occupational Health, Helsinki, Finland
R. Merad, Université d'Alger, Algeria
T.J. Meredith, Guy's Hospital, London, United Kingdom
E. Ngaha, Polytechnic of East London, London
Wai-On Phoon, National Occupational Health & Safety Commission, Sydney, Australia
Jenny Pronczuk de Garbino, CIAT, Montevideo, Uruguay
J. Rantanen, Institute of Occupational Health, Helsinki, Finland
Lynne Shaw, University of Reading, Reading, United Kingdom
E.M.B. Smith, International Programme on Chemical Safety, Geneva, Switzerland
G. Thiers, Instituté d'Hygiène et Epidémiologie, Brussels, Belgium
H.A.M.G. Vaessen, National Institute of Public Health & Environmental Protection, Bilthoven, The Netherlands
M.D. Watt, BDH Limited, Poole, United Kingdom
R. Wennig, Laboratoire National de Santé, Luxembourg
R. Wester, National Institute of Public Health & Environmental Protection, Bilthoven, The Netherlands
C.H. Williams, Jencons (Scientific) Limited, Leighton Buzzard, United Kingdom
Heather Wiseman, Poisons Unit, New Cross Hospital, London, United Kingdom
F. A. de Wolff, Academisch Ziekenhuis, Leiden, The Netherlands
E. Yärjanheikki, Institute of Occupational Health, Helsinki, Finland

Units, Nomenclature and Abbreviations

The use of units recognised in the Système International (SI) has been adopted throughout this monograph. Thus dm^3 and cm^3 are used instead of litre (L) and millilitre (mL) respectively. Various other quantities are expressed in the following units:

Temperature	°C
Pressure	Pa
Rate of flow	m/s
Molecular concentration	mol/dm^3 (preferred to "normality")
Low values of mass concentration	mg/kg (preferred to "ppm")
Heat of combustion	KJ/kg

In general, IUPAC nomenclature has been adopted for inorganic compounds, in preference to the trivial names listed here:

IUPAC nomenclature	Trivial name
Arsenate(III)	Arsenite
Arsenate(V)	Arsenate
Boron(III) fluoride	Boron trifluoride
Bromate(V)	Bromate
Chlorate(I)	Hypochlorite
Chlorate(III)	Chlorite
Chlorate(V)	Chlorate
Chlorate(VII)	Perchlorate
Chromate(VI)	(Di)chromate
Chromium(VI) oxide	Chromium trioxide
Iodate(V)	Iodate
Iodate(VII)	Periodate
Iron(II) sulfate	Ferrous sulfate
Manganate(VII)	Permanganate
Molybdate(VI)	Molybdate

Molybdenum(IV) sulfide	Molybdenum disulfide
Nitrate(III)	Nitrite
Nitrate(IV)	Nitrate
Phosphate(I)	Hypophosphate
Phosphate(III)	Phosphite
Phosphate(V)	Phosphate
Phosphorus(III) oxide	Phosphorus trioxide
Phosphorus(V) oxide	Phosphorus pentoxide
Sulfate(IV)	Sulfite
Sulfate(VI)	Sulfate
Sulfate(VII)	Persulfate

For organic compounds, trivial names are used in preference to systematic IUPAC nomenclature.

Abbreviations are generally those of standard convention:

b.p.	=	Boiling point
CCBW	=	Chemically contaminated biological waste
DFP	=	Diisopropyl fluorophosphate
HEPA	=	High efficiency particulate air filter
LEL	=	Lower explosive (flammability) limit *(see Glossary)*
LD_{50}	=	Lethal dose to 50% of a test group of a given species when administered orally or by some other route in a single dose
2-PAM	=	*N*-Methylpyridinium-2-aldoxime chloride (pralidoxime chloride)
SCBA	=	Self-contained breathing apparatus
TLV	=	Threshold limit value
TNT	=	Trinitrotoluene
UEL	=	Upper explosive (flammability) limit *(see Glossary)*
v.p.	=	Vapour pressure

Introduction

1. **NATURE AND SCOPE OF LABORATORY HAZARDS**

People who work in scientific laboratories are exposed to many kinds of hazards. Although this can be said of many kinds of workplaces, laboratories have a greater variety of hazards than most workplaces, and some of those hazards are seldom encountered elsewhere. In particular, laboratories in which chemicals are used must be prepared to deal with substances known to be hazardous, with new substances of unknown hazard, and with new kinds of experiments. In contrast to chemical manufacturing plants, laboratories usually handle only small amounts of materials, and exposure to a particular material seldom extends over a protracted time. In these respects, there is little difference among industrial, academic, and governmental laboratories. Colleges and universities must administer not only research and analytical laboratories, but also teaching laboratories, where inexperienced students must be introduced to the safety precautions needed for laboratory operations.

The past 40 years have witnessed extraordinary changes in the conduct of chemical operations, particularly in the ability to work with ever smaller samples and the use of automated instrumentation. Furthermore, our understanding of environmental and health problems has increased steadily, and is being reflected in improvements in laboratory design. The hazards of handling chemicals in the laboratory may be classified broadly as physical or chemical. Physical hazards include the familiar trio of fire, explosion, and electric shock. Other physical hazards arise from cryogenic equipment, refrigerators, furnaces, and means of containment, such as cylinders of compressed gases and glass apparatus.

Chemical hazards are associated with toxic effects and may be subclassified as acute or chronic. Acute hazards are those capable of producing prompt or only slightly delayed effects, such as burns, inflammation, allergic responses, or damage to the eyes, lungs, or nervous system. Some chemicals are extremely dangerous in this respect, and small amounts can cause death or severe injury very quickly. Some, such as chlorine or ammonia, have a warning odour; others, such as carbon monoxide, are not readily detected and are consequently more insidious.

Toxic effects of chemicals that are delayed or that develop only after exposure over long periods of time are referred to as chronic. Many organs can be damaged by cumulative chronic effects. Some effects are reversed on elimination of exposure to the chemical, but some are nearly irreversible, especially after much damage has occurred. Carcinogenicity is a notorious chronic effect.

In recent years, scientists and lay people alike have become increasingly conscious of the variety and seriousness of chronic effects, which are insidious because of the long delay in their appearance. However, the justifiable concern about chronic effects must not blind laboratory workers to the serious hazards of acutely toxic, explosive, and flammable substances, particularly since precautions against the latter hazards also reduce the probability of incurring chronic effects.

Much chemical research is concerned with new molecular structures, and it is common to synthesize substances that have never existed. When dealing with new structures, the laboratory worker should try to anticipate, by analogy to related structures of known toxicity, when a new substance may have high acute or chronic toxicity. However, predictions of biological activity are notoriously unreliable, and until actual toxicological data are available, all new substances should be handled as though toxic.

This book is concerned not only with the hazards that are encountered in the laboratory itself, but also with the related hazards involved in handling laboratory chemicals on the loading dock and in storerooms and stockrooms, in transporting them, and particularly in disposing of them. Precautions must be taken to protect the many other persons who may be exposed to chemical hazards arising from laboratory operations. These persons include those who enter the laboratory infrequently, such as maintenance personnel, and those who may be exposed to hazards in case of an accident or during transportation.

2. REGULATION OF USE OF CHEMICALS

Many types of regulations, differing in detail from country to country, affect the use of chemicals in the laboratory. The allowable exposures to compounds or classes of compounds in workplaces may be limited by law. The disposal of chemicals into sewers, landfills, or the atmosphere is often regulated, as is disposal by incineration. Transportation of chemicals and chemical waste is also commonly regulated. Those in charge of laboratories must know enough about the pertinent regulations to be in compliance with them. They must also ensure that all those working in the laboratory know enough about the regulations to carry out their operations safely and lawfully.

3. GOALS OF A LABORATORY SAFETY PROGRAMME

The principal objective of this book is to provide guidance to all laboratory workers who use chemicals so that they can perform their work safely. Experience has shown that the laboratory can be a safe workplace. This record, however, has been achieved only through vigorous safety programmes. The goals of a laboratory safety plan should be to protect those working in the laboratory, others who may be exposed to hazards from the laboratory, and the environment.

Chemicals must be disposed of in such a way that people, other living organisms, and the environment are subjected to minimal harm by the substances used or produced in the laboratory. Both laboratory workers and the supporting personnel should know and use acceptable disposal methods for various chemicals.

4. RESPONSIBILITY FOR LABORATORY SAFETY

First and foremost, the protection of health and safety in all laboratory operations, including waste-disposal, is a moral obligation. The expanding array of laws and regulations makes it also an economic necessity, and in many countries a legal requirement. In the final analysis, laboratory safety can be achieved only by the exercise of judgment by informed, responsible individuals. It is an essential part of the development of scientists that they learn to work with and to accept the responsibility for the appropriate use of hazardous substances.

Past experience has shown that voluntary safety programmes are usually inadequate. Good laboratory practice requires mandatory safety rules and programmes. To achieve safe conditions for the laboratory worker, a programme must include:

1. Regular safety inspections at intervals of no more than 3 months, and at shorter intervals for certain types of equipment such as safety showers and fire extinguishers.
2. Formal and regular safety programmes that ensure that at least some of the full-time personnel are trained in the proper use of emergency equipment and procedures
3. Regular monitoring of the performance of ventilation systems.
4. Disposal procedures that ensure proper disposal of waste chemicals at regular intervals.

A sound safety organization that is respected by all is essential; a good laboratory safety programme must always be based on participation of both laboratory administration and students and/or employees. A safety-minded point of view among all students and employees must be inculcated by the institution. In the end, the individual worker must learn to think about possible hazards and to seek information and advice before beginning any unfamiliar experiment.

The ultimate responsibility for safety within an institution lies with its chief executive officer. This individual must ensure that an effective institutional safety programme is in place. The chief executive officer and all immediate associates (e.g., vice presidents, deans, department heads) must have a continuing interest in the safety programme, and this interest must be obvious to everyone. An excellent safety programme that is ignored by top management will almost certainly be ignored by everyone else. An institutional safety coordinator is essential to an effective institutional safety programme. This individual should have had appropriate training and be qualified in those areas of safety that are relevant to the activities of the institution. Records should show that the facilities and the precautions taken in carrying out activities of an institution are compatible with current knowledge of the potential risks and with the law.

The responsibility for safety in a department lies with its chairperson or supervisor. A departmental safety coordinator is essential to an effective departmental safety programme. In smaller institutions, it may be possible for one person to perform more than one set of duties. For example, a significant fraction of the time of a departmental faculty member might be allotted to the duties of departmental safety coordinator.

The responsibility for the authorization of a specific operation, the delineation of the appropriate safety procedures, and the instruction of those who will carry out the operation lies with the project director. Although the responsibility for safety during the execution of an operation lies with those executing it, including scientists, technicians, other workers, and students, the primary responsibility still remains with the project director.

An effective departmental safety coordinator must be committed to the attainment of a high level of safety and must work with administrators and investigators to develop and implement written policies and practices appropriate for safe laboratory work. In these activities, the safety coordinator requires the cooperation of everyone - scientists, workers, technicians, and students.

Collectively, this group must routinely monitor current operations and practices, see that appropriate audits are maintained, and seek ways to improve the safety programme. If the goals of the laboratory dictate specific operations and the use of specific substances that are not appropriate to the existing facilities, it is the responsibility of the safety coordinator and the representative group to assist the investigator in acquiring adequate facilities and developing appropriate guidelines.

All accidents and near accidents should be carefully analysed, and the results of such analyses and the recommendations for the prevention of similar occurrences should be distributed to all who might benefit. This analysis is to be aimed at contributing to a safer environment, not at fixing blame.

5. HANDLING CHEMICALS IN THE LABORATORY

Chemicals occur in almost limitless, ever-increasing varieties. For this reason, general precautions suitable for handling almost all chemicals are needed, rather than specific guidelines for each chemical. Otherwise, laboratory work will be needlessly handicapped, practically and economically, by attempts to adhere to a labyrinth of specific guidelines. Under some circumstances, any chemical can be hazardous. Accordingly, this book recommends general procedures, applicable to all chemicals, that are designed to minimize the exposure of the laboratory worker and others to any chemical. In addition, it notes other precautions that are appropriate in working with substances that are known to be flammable, explosive, or unusually toxic.

It is imperative that hazards of the work in teaching and experimental research laboratories be differentiated from those in pilot-plants and industrial manufacture. Research in academic and industrial laboratories is carried out on a small-scale and hence generally involves low levels of exposure of laboratory workers to chemicals. This is particularly true when the laboratory worker makes proper use of the hoods (fume cupboards), protective apparel, and other safety devices that should be present in a well-equipped laboratory. Furthermore, in contrast to the typical industrial plant, where workers may be exposed to a limited number of substances over very long periods, the research worker is usually exposed to a large variety of substances at low levels for brief periods of time. Finally, through professional expertise, training, commonsense, judgment, and safety awareness, the research worker performing chemical operations in the laboratory is most often in the best position to judge necessary safety precautions.

5.1. Safety plans

Organizations that administer laboratories should have written safety plans available to, and understood by, all those working in the laboratory. In many organizations, safety committees representing the various types of persons exposed to potential hazards are needed to discuss problems, recommend safety measures, and facilitate communications.

Most organizations should have safety coordinators to serve as consultants on safety matters.

Educational guidance should be provided for all persons who may be exposed to potential hazards in connection with laboratory operations. This group includes faculty or teaching staff, students, laboratory supervisors, laboratory workers, maintenance and storeroom personnel, and others who are close to laboratories. New persons coming into the laboratory on related jobs should be instructed in safety procedures and the procedures to use in the event of accidents. The safety programme, including educational activities, should be a regular, continuing effort and not merely a stand-by activity that functions for a short time after each laboratory accident.

5.2. Medical programme

Any person whose work involves regular and frequent handling of toxicologically significant quantities of material that is acutely or chronically toxic should be encouraged by the employer to consult a physician qualified to deal with toxicological problems so as to determine whether it is desirable to establish a regular schedule of medical surveillance. For example, it is important to monitor body concentrations of cumulative poisons such as lead or mercury. In some countries, biological monitoring is mandated by law, as part of the medical surveillance procedure.

5.3. Facilities

Laboratory facilities include appropriate ventilation and fume extraction systems, stockrooms and storerooms, safety equipment, and arrangements for disposal of chemical waste. Although the energy costs of ventilation, often substantial, are increasing, considerations of economy should never take precedence over ensuring that laboratories have adequate ventilation. Any change in the overall ventilation system to conserve energy should be instituted only after thorough testing of its effects has demonstrated that the laboratory workers will continue to have adequate protection from hazardous concentrations of airborne toxic substances. An inadequate ventilation system can cause increased risk because it may give rise to a false sense of security.

5.4. Monitoring of chemical substances

For most laboratory environments, regular monitoring of the airborne concentrations of a variety of different toxic materials is unnecessary. If care is taken to ensure that the fume extraction and the rest of the ventilation system are performing efficiently and being used properly, that the laboratory workers are using proper protective clothing to avoid skin contact, and are following other good practices, then even highly toxic materials can be handled without undue hazard.

There are two circumstances where monitoring of individual compounds is appropriate. First, if the fume extraction or other local ventilation devices in a laboratory are being tested or redesigned, it is desirable to check their performance by releasing in the laboratory a small amount of a relatively non-toxic substance (e.g., fluorotrichloromethane or sulfur hexafluoride) whose airborne concentration is readily monitored by commercial instruments. Alternatively, laboratory workers can wear personal air-sampling devices to provide a measure of the airborne concentration of some substance in their environment. Second, if a highly toxic substance is being used regularly, instrumental monitoring for that substance may be appropriate.

5.5. Academic teaching laboratories

In general, the students and instructors in academic teaching laboratories should follow the safety procedures recommended for full-time laboratory workers in research and development laboratories. The most severe limitations on protective equipment in instructional laboratories are usually the general laboratory ventilation and auxiliary local exhaust ventilation such as hoods. Unless adequate hood space can be provided, it is prudent to avoid work with substances whose toxicity is not known. The selection of the particular substances to use among those whose toxicological properties are known should take into account the quality of the ventilation system available.

5.6. Literature and consultant advice

Literature (see Bibliography) and consultant advice on laboratory safety and on the physical and biological hazards of chemicals should be readily available to those responsible for laboratory operations and those actually involved.

Although substantial numbers of people who have expertise in laboratory safety are employed by large chemical companies and by private consulting firms, such persons are not often found in academic institutions. Because modifications of certain safety facilities (e.g., ventilation systems, waste-disposal systems) can be very expensive, such modifications should not be undertaken until advice has been sought from persons qualified to make recommendations. The alternatives for an academic institution are to hire either an appropriately qualified person as the institutional safety coordinator or an appropriate consultant, when required, for advice about a specific safety problem.

6. LABORATORY WASTE-DISPOSAL

All laboratory work with chemicals eventually produces chemical waste, and those who generate such waste have moral and, in most countries, legal obligations to see that the waste is handled and disposed of in ways that pose minimum potential harm, both short-term and long-term, to health and to the environment.

For the purposes of this book, a laboratory is defined as a building or area of a building used by scientists or engineers, or by students or technicians under qualified supervision, for the following purposes: investigation of physical, chemical, or biological properties of substances; development of new or improved chemical processes, products, or applications; analysis, testing, or quality control; or instruction and practice in a natural science or in engineering. These operations are characterized by the use of a relatively large and variable number of chemicals on a scale in which the containers used for reactions, transfers, and other handling of chemicals are normally small enough to be easily and safely manipulated by one person.

Broadly, a hazardous chemical is a chemical that poses a danger to human health or to the environment if improperly handled. Countries may define hazardous waste, for regulatory purposes, in terms of specific hazard characteristics and by listing specific chemicals and residues from chemical operations. Even though many of the chemicals synthesized or used in laboratories do not meet any such regulatory criteria, they must be considered hazardous because of unknown toxicity. This book is concerned with all laboratory chemicals that must be handled and disposed of as hazardous, regardless of whether they are covered by government regulations. Biological waste generated by life-science laboratories is covered only to the extent that the biological waste is contaminated with chemicals; for example, waste generated in animal or microbiological tests of the toxicity or mode of action of chemicals. This book does not cover radioactive waste, which is specifically regulated almost everywhere by national laws.

A large fraction of laboratory waste comprises small amounts of many kinds of chemicals; such waste is generally sent to a secure landfill, sometimes hundreds of kilometres from the laboratory, because many operators of commercial incinerators do not accept these low-volume, chemically diverse wastes. Because of these constraints on disposal, and the associated extensive record keeping required, many laboratories have problems in setting up effective systems for handling their hazardous waste and, in some cases, in finding legal and safe ways for disposing of it. These problems are discussed extensively in the book, but it is recognized that there is a steady stream of new legislation arising from public concern in many countries over waste disposal issues.

The characteristics and requirements of a waste-management system are outlined in Chapter 12. A clearly documented waste-management plan should be an important part of the written safety plan of a laboratory.

Many of the scientists being educated today in our colleges and universities will move into professions from which society expects responsible, informed behaviour in the disposal of hazardous chemical substances. The time to learn this behaviour is in the high school and college or university laboratory. Educators, many of whom were trained in a more casual era, must now be responsible, by precept and example, for teaching proper methods for the disposal of hazardous laboratory waste. Teachers should demand that a formal waste-disposal programme be initiated if it has not yet been, and they should see that the programme is implemented in their laboratories.

Undergraduate laboratories should be equipped with the necessary disposal equipment, and the proper disposal of wastes should be a part of every laboratory procedure. The authors of laboratory texts should deal with the problems of disposal explicitly and should include as part of each experiment procedures that the student can carry out to minimize disposal problems. Laboratory procedures for destroying the hazardous characteristics of many classes of common laboratory chemicals are given in Chapter 10. In some academic institutions, the safe disposal procedures for hazardous laboratory chemicals can be incorporated into the curriculum.

Professionals trained in this area will be in demand, and therefore programmes on disposal of chemicals from laboratories can be incorporated into a wide range of training programmes. Chemistry students can learn about detoxifying or destroying chemicals, micobiology students can come to grips with aerobic and anaerobic destruction of toxic materials, and engineering students can study design of incinerators and other waste-handling equipment.

Part I

Safe Working Procedures and Protective Equipment

1

General Recommendations for Safe Practices in Laboratories

It is impossible to design a set of rules that will cover all possible laboratory hazards and occurrences. The general guidelines given below are based largely on experience and have helped to avoid accidents and reduce injuries in laboratories.

Safe working is the most important rule that all laboratory personnel must be aware of, from the senior administrator to the laboratory aid. Safety awareness becomes part of each worker's habit if the subject is discussed repeatedly, and if senior staff exhibit a genuine and continued interest. The individual, however, must accept responsibility for carrying out laboratory work in accordance with good safety practices, and should be prepared in advance for possible accidents by knowing what emergency aids are available and how they are to be used.

The supervisor of the laboratory has overall responsibility for safety, and should provide for regular formal safety and housekeeping inspections (at least quarterly for universities and other organizations that have frequent changes of personnel, and six monthly for other laboratories) in addition to continuing informal inspections. Laboratory supervisors have the responsibility of ensuring that:

1. Workers know safety rules and follow them.
2. Adequate emergency equipment is available in full working order.
3. Training in the use of emergency equipment has been provided.
4. Information on special or unusual hazards in non-routine work has been distributed to the laboratory workers.
5. Appropriate safety training has been given to individuals when they commenced their work in the laboratory.

The laboratory worker should develop good personal safety habits:

1. Appropriate eye protection should be worn at all times.
2. Exposure to chemicals should be kept to a minimum.
3. Smoking, eating, drinking and the application of cosmetics should be prohibited in areas where laboratory chemicals are present.

Advance planning is one of the best ways to avoid serious incidents. Before performing any chemical operation, the laboratory worker should consider "What would happen if...?" and be prepared to take proper emergency actions.

Overfamiliarity with a particular laboratory operation may result in overlooking or underrating its hazards. This attitude can lead to a false sense of security, which frequently results in carelessness. Every laboratory worker has a basic responsibility to plan and execute laboratory procedures in a safe manner.

1.1. GENERAL PRINCIPLES

Every laboratory worker should observe the following rules:

1. Know the safety rules and procedures that apply to the work that is being carried out. Determine the potential hazards (e.g., physical, chemical, biological - see Chapters 6-8) and appropriate safety precautions *before* beginning any new operation.
2. Know the types of protective apparel available and use the proper type for each job (Sections 2.1-2.3 of chapter 2).
3. Know the location of the safety equipment in the area, as well as how to use it and how to obtain additional help in an emergency (Section 2.4). Be familiar with emergency and first aid procedures (Sections 2.5-2.6).
4. Be alert to unsafe conditions and actions, and call attention to them so that corrections can be made as soon as possible. Someone else's accident can be as dangerous to you as one of your own.
5. Prohibit smoking, the consumption of food or beverages and the application of cosmetics in areas where laboratory chemicals are being used or stored (Section 1.3).
6. Prevent hazards to the environment by following accepted waste-disposal procedures. Chemical reactions may require traps or scrubbing devices to prevent the escape of toxic substances.

7. Be certain that all chemicals are correctly and clearly labelled. Combine reagents in the appropriate order.
8. Post warning signs when unusual hazards, such as radiation, laser operations, flammable materials, biological hazards or other special problems exist.
9. Remain out of the area of a fire or personal injury unless it is your responsibility to help meet the emergency. Curious bystanders may interfere with rescue and emergency personnel and endanger themselves (Section 2.5).
10. Avoid distracting or startling any other worker. Practical jokes or horseplay cannot be tolerated in the laboratory.
11. Use equipment only for its designed purpose.
12. Position and clamp reaction apparatus thoughtfully in order to permit manipulation without the need to move the apparatus until the reaction is completed.
13. Think, act, and encourage safety until it becomes a habit.
14. Participate in a continuous safety training programme.

1.2. HEALTH AND HYGIENE

Laboratory workers must observe the following health practices:

1. Wear appropriate eye protection at all times (Section 2.1).
2. Use protective apparel, including face shields, gloves, and other special clothing or footwear as needed (Sections 2.2, 2.3).
3. Confine long hair and loose clothing when in the laboratory (Section 2.3).
4. Do not use mouth suction to pipette chemicals or to start a siphon; a pipette bulb or an aspirator should be used.
5. Avoid exposure to gases, vapours, and aerosols (Chapter 3). Use appropriate safety equipment whenever such exposure is likely.
6. Wash well before leaving the laboratory area. However, avoid the use of solvents for washing skin. They remove natural protective oils and may cause irritation / inflammation; moreover, they can facilitate absorption of chemicals through the skin.

1.3. FOOD HANDLING

Contamination of food, drink, and smoking materials is a potential route for exposure to toxic substances. Food should not be stored, handled, or consumed in any laboratory area.

1. Well-defined areas should be established for storage and consumption of food and beverages. No food should be stored or consumed outside this area.

2. Areas where food is permitted should be prominently marked and a warning sign (e.g., EATING AREA - NO CHEMICALS) displayed. All chemicals, chemical equipment and laboratory coats must be excluded from such areas.
3. Consumption of food or beverages and smoking must be excluded from areas where laboratory operations are carried out.
4. Glassware or utensils that have been used for laboratory operations should never be used to prepare food or beverages. Laboratory refrigerators, ice chests, and cold rooms should not be used for food storage; separate equipment should be dedicated for that use and prominently labelled.

1.4. HOUSEKEEPING

There is a definite relationship between safety performance and orderliness in the laboratory. When housekeeping standards fall, safety performance inevitably deteriorates. The work area should be kept clean, and chemicals and equipment should be properly labelled and stored.

1. Work areas should be kept clean and free from obstructions. Cleaning up should follow the completion of any operation or take place at the end of each day.
2. Wastes should be deposited in appropriate colour coded or labelled receptacles.
3. Spilled chemicals should be cleaned up immediately and disposed of properly. Disposal procedures should be established, and all laboratory personnel should be aware of them (Part III); the effects of other laboratory accidents should also be cleaned up promptly.
4. Unlabelled containers and chemical wastes should be disposed of promptly by appropriate procedures (Part III). Such materials, as well as chemicals that are no longer needed, should not be accumulated in the laboratory.
5. Floors should be cleaned regularly; accumulated dust, chromatography adsorbents, and various chemical residues may pose respiratory hazards.
6. Corridors, hallways and stairways should not be used as storage areas, either for chemicals or equipment of any kind.
7. Access to exits, emergency equipment and switches or control valves should never be blocked.
8. Equipment and chemicals should be stored properly.

1.5. EQUIPMENT MAINTENANCE

Good equipment maintenance is important for safe, efficient operations. Equipment should be inspected and maintained regularly.

Servicing schedules should be related to expected usage and the reliability of the equipment. Equipment awaiting repair or maintenance should be removed from service.

1.6. GUARDING FOR SAFETY

All mechanical equipment should be furnished with guards that prevent access to electrical connections or moving parts. Each laboratory worker should inspect equipment before using it to ensure that the guards are in place.

Careful design of guards is vital. An ineffective guard can be worse than none at all as it can give rise to a false sense of security. Emergency shut-off devices may be needed, in addition to electrical and mechanical guarding.

1.7. SHIELDING FOR SAFETY

Safety shielding should be used for any operation that is potentially explosive, such as:

1. An exothermic reaction being attempted for the first time.
2. A familiar reaction carried out on a larger-than-usual scale.
3. Operations being carried out under non-ambient pressures (Sections 8.2, 8.3). Shields must be placed so that all personnel in the area are protected from hazard.

1.8. GLASSWARE

Accidents involving glassware are a major cause of laboratory injuries. These can be avoided or reduced by:

1. Careful handling and storage procedures should be used to avoid damaging glassware. Damaged items should be discarded or repaired by an experienced glassblower.
2. Glassblowing operations should not be attempted except at a properly equipped bench, with proper annealing facilities being available.
3. Vacuum-jacketed glass apparatus should be handled with extreme care to prevent implosion. Equipment such as Dewar flasks should be taped or shielded (Section 8.3). Only glassware designed for vacuum work should be used for that purpose.
4. Thick gloves should be worn when picking up pieces of broken glass. Small fragments should be swept up with a brush into a dustpan.
5. Proper instruction should be provided in the use of glass equipment designed for special tasks which might present unusual risks. For example, stoppered separating funnels containing volatile solvents can develop considerable pressure.

6. Exceptionally, it may be necessary to insert glass tubing into bungs or corks, or make rubber-to-glass hose connections. In these cases tubing should be fire polished and lubricated, and appropriate hand protection should be worn.

1.9. FLAMMABILITY HAZARDS

Because flammable materials are widely used in laboratory operations (Chapter 7), the following rules should be observed:

1. Store flammable materials correctly (Section 5.4).
2. Never use an open flame to heat a flammable liquid or to carry out a distillation under reduced pressure.
3. In any laboratory procedure an open flame should only be used when there is no better alternative. It should be extinguished once it is no longer required.
4. Before lighting a flame, remove all flammable substances from the immediate area. Check all containers of flammable materials in the area to ensure that they are tightly closed.
5. Notify other occupants of the laboratory before lighting a flame.
6. When volatile flammable materials may be present, use only intrinsically safe non-sparking electrical equipment, which is not liable to over-heat.

1.10. COLD TRAPS AND CRYOGENIC HAZARDS

Cryogenic materials and the surfaces they cool can cause severe "cold" burns if they are allowed to contact the skin. Thick gloves and a face shield may be needed when preparing or using some cold baths (Sections 2.1, 2.2).

Neither liquid nitrogen nor liquid air should be used to cool a flammable mixture in the presence of air because both can condense oxygen from the air, resulting in a potentially explosive mixture. "Dry ice" (solid carbon dioxide) should be handled with dry leather gloves, and should be added slowly to the liquid cooling bath to avoid foaming. Workers should avoid lowering their heads into a "dry ice" chest; a high concentration of carbon dioxide can inhibit the autonomic breathing system, which can lead to loss of consciousness or asphyxiation. Large quantities of liquid nitrogen can also cause asphyxiation in confined or poorly ventilated spaces.

1.11. CLOSED-SYSTEMS UNDER POSITIVE OR NEGATIVE PRESSURE

Reactions should never be carried out in, or heat applied to, a closed-system unless it is designed and tested to withstand the expected pressure change. Pressurized apparatus should have an appropriate relief device.

1.12. WASTE-DISPOSAL PROCEDURES

Laboratory managers have the responsibility for establishing waste-disposal procedures for routine and emergency situations (Chapter 12) and for communicating these procedures to laboratory workers. Workers must follow them to avoid hazards or damage to the environment.

1.13. WARNING SIGNS AND LABELS

Laboratory areas that have special or unusual hazards should be posted with warning signs. Standard signs and symbols have been established for a number of special situations, such as radioactivity hazards, biological hazards, fire hazards, and laser operations. Other signs should be posted to show the locations of safety showers, eyewash stations, fire extinguishers and exits. Extinguishers should be sited thoughtfully and they should be clearly labelled showing the type of fire for which they are intended (Section 2.4.1). Waste containers should be colour coded or labelled indicating the type of waste that can be safely deposited in them. The safety and hazard signs in the laboratory should enable personnel to deal appropriately with an emergency situation.

When possible, labels on chemical containers should contain information concerning the hazard(s) associated with use of the chemical. Unlabelled bottles should not be allowed to accumulate. They should be handled with care while determining how they can be disposed of safely (Section 10.4).

1.14. UNATTENDED OPERATIONS

Laboratory operations are often carried out continually or overnight with no one present. It is essential to plan for interruptions in utility services such as electricity, water, and gases. Operations should be designed to be safe, and plans should be made to avoid hazards in case of failure. Wherever possible, arrangements for periodic inspection of the operation should be made and, in all cases, the area should be properly lit and an appropriate sign should be placed on the door and the apparatus.

Failure of cooling water can have serious consequences. A variety of devices can be used that:

1. Automatically regulate water pressure to avoid surges that might rupture the supply lines, or
2. Monitor the water flow so that a failure will automatically result in the turning off of electrical connections and water supply valves, and vent any pressure build-up in the system.

1.15. WORKING ALONE

It is prudent to avoid working in a laboratory alone. If this must be done, however, arrangements should be made between individuals working in separate laboratories outside of conventional hours to cross-check periodically. Alternatively, security guards may be asked to check on a laboratory worker. Experiments known to be hazardous should not be undertaken by a worker who is alone in a laboratory.

Special rules may be necessary under unusual circumstances. The supervisor of the laboratory has the responsibility for determining whether the work requires special safety precautions, such as having two persons in the same room or in close proximity during a particular operation.

1.16. ACCIDENT REPORTING

Emergency telephone numbers to be used in the event of fire, accident, flood, or hazardous chemical spillage should be displayed prominently in each laboratory. In addition, the home telephone numbers of the laboratory workers and their supervisors should be displayed. The appropriate persons must be notified promptly in the event of an accident or emergency.

Every laboratory should have an internal accident reporting system. This includes provisions for investigating the cause of an injury as well as any potentially serious incident that does not result in injury. The primary aim of such investigations should be to make recommendations to improve safety, not to assign blame for an incident. Local legal regulations may require reporting procedures for accidents or injuries.

1.17. EVACUATION

Not only should laboratory workers be aware of emergency telephone numbers, but they should know the location and proper use of fire fighting appliances, rescue equipment and emergency exits. They should also be familiar with the prescribed procedure for evacuating the building, and the location of the outside assembly point where personnel rolls will be checked.

1.18. EVERYDAY HAZARDS

Laboratory workers should remember that injuries can and do occur outside the laboratory in other work areas. It is important that safety principles be practiced in offices, stairways, corridors, and similar places. Here, safety is largely a matter of common sense, but a constant awareness of everyday hazards is vital.

2

Protective Apparel, Safety Equipment, Emergency Procedures and First Aid

A wide variety of specialized clothing and equipment is commercially available for use in the laboratory. The proper use of these items minimizes or eliminates exposure to the hazards associated with many laboratory operations. The primary goal of laboratory safety procedures is the prevention of accidents and emergencies. However, they may nonetheless occur, and at such times, proper safety equipment and correct emergency procedures can help to minimize injury or damage.

2.1. SPECTACLES AND FACE SHIELDS
2.1.1. General eye protection policy

Eye protection must be required for all personnel, and any visitors, in locations where chemicals are stored or handled. No one should enter any laboratory without appropriate eye protection.

Safety glasses that meet the criteria described below provide minimum eye protection for regular use. Additional protection may be required when carrying out more hazardous operations.

Contact lenses should not be worn in a laboratory. Gases and vapours can be concentrated under the lenses and cause permanent damage. Furthermore, in the event of a chemical splash into an eye, it is often extremely difficult to remove the contact lens to irrigate the eye because of involuntary spasm of the eyelid. Thus when irrigating the eyes of an unconscious person the presence of contact lenses will reduce the effectiveness of such treatment. Soft lenses can absorb solvent vapours even through face shields and, as a result, adhere to the eye.

There are some situations in which contact lenses must be worn for therapeutic reasons. Persons who must wear contact lenses should inform the laboratory supervisor so that satisfactory safety precautions can be devised.

2.1.2. Safety spectacles

Ordinary prescription spectacles do not provide adequate protection against eye injury. The minimum acceptable eye protection requires the use of hardened-glass or plastic safety glasses with a minimum lens thickness of 3 mm, good impact and flammability resistance, and lens-retaining frames. Cellulose nitrate frames should never be used.

Side shields that attach to regular safety glasses offer some protection from objects that approach from the side but do not provide adequate protection from splashes. Other eye protection should be worn when a significant splashing hazard exists.

2.1.3. Other eye protection

Other forms of eye protection that may be required for a particular operation include the following:

Goggles

These are not intended for general use. They are intended for wear when there is danger of splashing chemicals or flying particles. For example, goggles should be worn when working with glassware under reduced or elevated pressure, and when glass apparatus is used in combustion or other high temperature operations. Impact protection goggles have screened areas on the sides to provide ventilation and reduce fogging of the lens, so they do not offer full protection against chemical splashes. Splash goggles ("acid goggles") that have splash-proof sides should be used when protection from harmful chemical splashes is needed.

Face shields

Goggles offer little protection to the face and neck. Full-face shields that protect the face and throat should always be worn when maximum protection from flying particles and sprays of harmful liquids is needed. For full protection, safety glasses should be worn with face shields. A face shield or mask may be needed when a vacuum system (which may implode) is used, or when a reaction has potential for explosion.

Specialized eye protection

There are specific goggles and masks for protection against laser hazards and ultraviolet or other intense light sources, as well as for glassblowing and welding operations.

2.2. GLOVES

Gloves should be worn whenever it is necessary to handle corrosive materials, rough or sharp-edged objects, very hot or very cold materials, or whenever protection is needed against accidental exposure to chemicals. Gloves should not be worn around moving machinery.

Many different types of gloves are commercially available.

1. Leather gloves may be used for handling broken glassware, for inserting glass tubes into rubber stoppers, and for other operations where protection from chemicals is not needed.
2. There are various compositions and thicknesses of rubber gloves. Common glove materials include neoprene, polyvinyl chloride and nitrile, butyl, and natural rubbers. These materials differ in their resistance to various substances. Specific information is often available from glove manufacturers' catalogues; such information is given in Table 2.1. Rubber or plastic gloves should be inspected before each use. A periodic inflation test should be conducted, in which the glove is first inflated with air and then immersed in water and examined for the generation of air bubbles from pinholes.
3. Insulated gloves should be used when working at extreme temperatures. Various synthetic materials such as Nomex® and Kevlar® can be used briefly up to 540 °C. Gloves made of these materials or in combination with other materials such as leather are available. It is best not to use gloves made either entirely or partly of asbestos ,a recognized carcinogen, although such gloves probably do not present much hazard.

Skin contact is a potential source of exposure to toxic materials (Section 6.1.3.). It is important to take proper steps to prevent such contact, considering the following points:

1. Proper protective gloves should be worn whenever there is potential for contact with corrosive or toxic materials and materials of unknown toxicity.
2. Gloves should be selected on the basis of the substance(s) being handled, the particular hazard involved, and their suitability for manipulations in the operation being conducted.
3. Before each use, gloves should be inspected for punctures or tears, and other weaknesses that may be indicated by discolouration.
4. Before removal, gloves should be washed appropriately. Some gloves, e.g., leather and polyvinyl chloride, are water-permeable.
5. Glove materials are eventually permeated by chemicals. They should be replaced periodically, depending on frequency of use and permeability to the substances handled.
6. Many carcinogens either in the solid state or in solution can permeate glove material. It is advisable therefore to work with

Table 2.1. Resistance of common glove materials to chemicals

Chemical	Natural rubber	Neoprene	Nitrile rubber	Polyvinyl chloride
Acetaldehyde	G	G	E	G
Acetic acid	E	E	E	E
Acetone	G	G	G	F
Acrylonitrile	P	G	-	F
Ammonium hydroxide (29%)	G	E	E	E
Aniline	F	G	E	G
Benzaldehyde	F	F	E	G
Benzene[a]	P	F	G	F
Benzyl chloride[a]	F	P	G	P
Bromine	G	G	-	G
Butyraldehyde	P	G	-	G
Carbon disulfide	P	P	G	F
Carbon tetrachloride[a]	P	F	G	F
Chlorine	P	G	-	G
Chloroform[a]	P	F	G	P
Chromic(VI) acid	P	F	F	E
Cyclohexane	F	E	-	P
Dibutyl phthalate	F	G	-	P
1,2-Dichloroethane[a]	P	F	G	P
1,2-Dichloropropane[a]	P	F	-	P
Diethyl ether	F	G	E	P
Dimethyl sulfoxide[b]	-	-	-	-
Ethyl acetate	F	G	G	F
Ethylene glycol	G	G	E	E
Fluorine	G	G	-	G
Formaldehyde (37%)	G	E	E	E
Formic acid	G	E	E	E
Hexane	P	E	-	P
Hydrobromic acid (40%)	G	E	-	E
Hydrochloric acid (30-36%)	G	G	G	E
Hydrogen peroxide	G	G	G	E
2-Methoxyethanol	F	E	-	P
Methylene chloride[a]	F	F	G	F
Monoethanolamine	F	E	-	E
Morpholine	F	E	-	E
Nitric(V) acid (71%)	P	P	P	G
Chloric(VII) acid	F	G	F	E
Phenol	G	E	-	E
Sodium hydroxide (10%)	G	G	G	E
Sodium chlorate(I) (5%)	G	P	F	G
Sulfuric acid (96%)	G	G	F	G

Table 2.1 Continued

Chemical	Natural rubber	Neoprene	Nitrile rubber	Polyvinyl chloride
Trichloroethylene[a]	P	F	G	F
Tricresyl phosphate	P	F	-	F
Trinitrotoluene	P	E	-	P

(E = Excellent, G = Good, F = Fair, P = Poor)

[a] Aromatic and halogenated hydrocarbons will attack all types of natural and synthetic glove materials. Should swelling occur, the user should change to fresh gloves.

[b] No data are available on the resistance to dimethyl sulfoxide of natural rubber, neoprene, nitrile rubber, or vinyl materials. A manufacturer of dimethyl sulfoxide recommends the use of butyl rubber gloves.

two pairs, replacing the outer pair as soon as any contamination is observed.
7. Some carcinogens have electrostatic properties which, if they are handled by latex or vinyl gloves, cause increased dissemination into the atmosphere. This may be prevented by the use of disposable cotton gloves.

2.3. OTHER CLOTHING AND FOOTWEAR

The everyday clothing worn by laboratory workers can be important to their safety. They should not wear loose clothing (e.g., saris, dangling neckties, overlarge laboratory coats), skimpy clothing (e.g., shorts, halter tops), torn clothing, or unrestrained long hair. Loose or torn clothing and unrestrained long hair can easily catch fire, dip into chemicals, or become ensnarled in apparatus and moving machinery; skimpy clothing offers little protection to the skin in the event of chemical splash.

Finger rings should be avoided when working with equipment that has moving parts. Shoes should be worn at all times in buildings where chemicals are stored or used. Perforated or cloth shoes, or sandals, should not be worn in laboratories.

2.3.1. Protective apparel

Protective apparel is advisable for most laboratory work and may be required for some. This includes laboratory coats and aprons, jump suits, special types of boots, shoe covers, and gauntlets which may be washable or disposable. These will help protect against chemical splashes or spills, heat, cold, moisture, and low-energy radiation.

Laboratory coats are intended to prevent contact with dirt and minor chemical splashes or spillage. The cloth laboratory coat is, however, primarily a protection for clothing and may itself present a hazard to the wearer (e.g., combustibility). Cotton and synthetic materials such as Nomex® or Tyvek® are generally satisfactory; rayon and polyesters are not. Laboratory coats provide little resistance to penetration by organic liquids and should be removed immediately if significantly contaminated by them.

Plastic or rubber aprons provide better protection from corrosive or irritating liquids, but can complicate injuries in the event of fire. Furthermore, a plastic apron can accumulate a considerable static charge and should be avoided in areas where flammable solvents or other materials could be ignited by a static discharge, or in the handling of carcinogens.

Disposable outer garments (e.g., Tyvek®) may sometimes be preferable to reusable ones. An example is the handling of very carcinogenic compounds (Section 6.9), where long sleeves and the use of gloves are also recommended. Disposable full length jump suits are strongly recommended for high risk situations, which may also require the use of head and shoe covers. Many disposable garments, however, offer only limited protection from vapour penetration, and judgment is needed in deciding when to use them. Impervious suits that fully enclose the body may be necessary in emergency situations.

Chemical spillage on leather clothing or accessories (e.g., watchbands, shoes, belts) can be especially hazardous because many chemicals are absorbed in the leather and then held close to the skin for long periods. Such items must be removed promptly and decontaminated or discarded to prevent the possibility of chemical burns, or allowing absorption through the skin.

2.3.2. Foot protection

More extensive foot protection than ordinary shoes may sometimes be required. Rubber boots or plastic shoe covers may be used to avoid exposure of the feet to corrosive chemicals or large quantities of solvents or water that might penetrate normal foot gear (e.g., during clean-up operations). Because these types of boots and covers may increase the risk of static spark, their use in normal laboratory operations is not advisable. Other specialized tasks may require footwear that has, for example, conductive soles, insulated soles, or built-in metal toe caps.

2.4. SAFETY EQUIPMENT

All laboratories in which chemicals are used should have available fire extinguishers, safety showers, and eyewash fountains, as well as fume extraction systems and laboratory sinks. The last item may be considered part of the safety equipment of the laboratory. Indeed, legislation in some countries requires the fitment of a minimum number of hand wash basins. Respiratory protection for emergency use should be readily available, along with fire alarms, emergency telephones, and identified emergency telephone numbers (Section 1.16).

2.4.1. Fire safety equipment

Fire safety equipment in the laboratory includes a variety of extinguishers together with hoses, blankets, and fire detection and automatic extinguishing systems.

Fire extinguishers

All chemical laboratories should be provided with carbon dioxide (Figure 1) or dry chemical (Figure 2) fire extinguishers or both. Other types should be available if required by the work being done. The four types of extinguishers most commonly used are classified by the type of fire for which they are suitable.

Fig. 1 Extinguishing a flaming liquid in a hood (fume cupboard) by directing carbon dioxide at the base of the fire.

Fig. 2 Extinguishing a fire in a hood (fume cupboard) with dry powder - carbonate or phosphate; this kind of extinguisher can be used effectively at a greater distance from a fire than a carbon dioxide extinguisher.

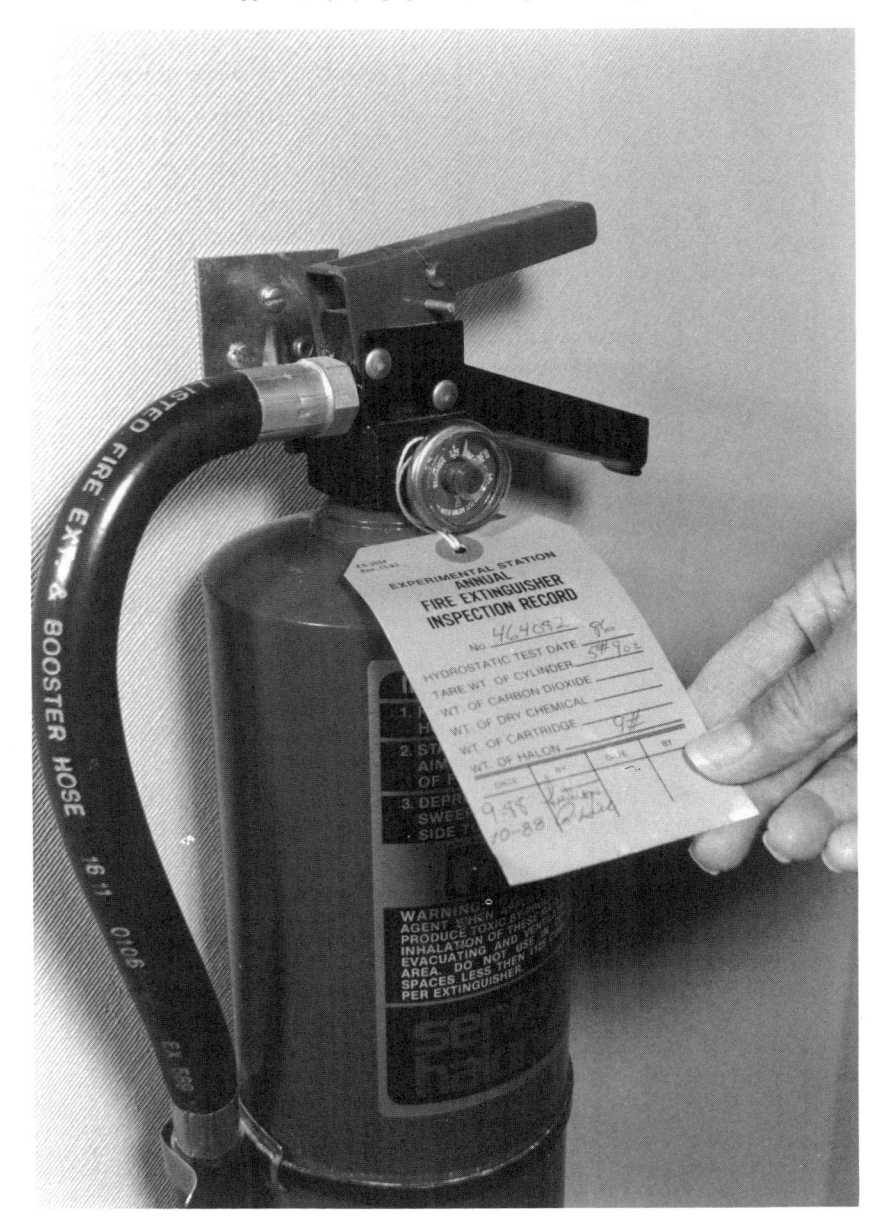

Fig. 3   Example of a modern type of wall-mounted fire extinguisher bearing a label showing the last date of inspection.

1. Water extinguishers are effective against burning paper and ordinary rubbish (Class A fires). They should not be used on electrical, liquid, or metal fires.
2. Carbon dioxide extinguishers are effective against burning liquids, such as hydrocarbons or paint, and electrical fires (Classes B and C fires). They are recommended for fires involving delicate instruments and optical systems because they do not damage such equipment. They are less effective against paper, rubbish, or metal fires and should not be used against lithium aluminium hydride fires.
3. Dry powder extinguishers, which contain sodium hydrogen carbonate, are effective against burning liquids and electrical fires (Classes B and C fires). They are less effective against paper, rubbish, or metal fires. They are not recommended for fires involving delicate instruments or optical systems because of the clean-up problem. These extinguishers are generally used where large quantities of solvent may be present. They are effective at a greater distance from a fire than carbon dioxide extinguishers.
4. Extinguishers that have special granular formulations are effective against burning metal (Class D fires). Included in this category are fires involving magnesium, lithium, sodium, and potassium, alloys of reactive metals, and metal hydrides, metal alkyls, and other organometallics. These extinguishers are less effective against paper, rubbish, liquid, or electrical fires.

Every extinguisher should carry a label indicating what class or classes of fire it is effective against. There are a number of other more specialized types of extinguishers for unusual fire hazards. Each laboratory worker should be responsible for knowing the location, operation, and limitations of the fire extinguishers in the work area. It is the responsibility of the laboratory supervisor to ensure that all workers are shown their locations and are trained in their use. Extinguishers should be checked regularly to ensure that they are full and in good condition (Figure 3). After use, an extinguisher must be recharged or replaced by designated personnel.

Fire hoses

Fire hoses are intended for use by trained firefighters against fires too large to be handled by extinguishers. Water has a cooling action and is effective against fires involving paper, wood, rags and rubbish (Class A fires). Water should not be used directly on fires that involve live electrical equipment (Class C fires) or chemicals such as alkali metals, metal hydrides, and metal alkyls that react vigorously with it (Class D fires).

Streams of water should not be used against fires that involve oils or other water insoluble flammable liquids (Class B fires). This form of water will not readily extinguish such fires, and it will usually spread or float the fire to adjacent areas. These possibilities are minimized by the use of a water fog.

Because of the potential hazards in using water around chemicals or electrical apparatus and because of the difficulty of controlling a hose delivering water at high pressure, laboratory workers should refrain from using fire hoses except in extreme emergencies. Their use should be reserved for trained firefighters.

Fire blankets

Many laboratories still have fire blankets available, even though their usefulness is limited. A fire blanket is used primarily as a first aid measure for the prevention of shock rather than against smouldering or burning clothing. It should be used only as a last resort to extinguish clothing fires, as blankets tend to hold heat in and increase the severity of burns. Clothing fires should be extinguished by immediately dropping to the floor and rolling or by using a safety shower, if it is close by the incident. A fire blanket, cautiously put into position, may be helpful in containing or even extinguishing a fire in a small open vessel.

Automatic detection and fire extinguishing systems

In areas of high fire potential and high risk of injury or damage (for example, solvent storage areas), automatic detection and fire extinguishing systems are often used. These may be of the water sprinkler, carbon dioxide, dry chemical, or halogenated hydrocarbon types. Where these are fitted, the laboratory workers should be informed of their presence and type, and be advised of any safety precautions required for their use (e.g., evacuation, before a carbon dioxide total flood system is actuated).

2.4.2. Respiratory protective equipment

The primary way to protect laboratory personnel from airborne contaminants is to minimize the amount of contaminants in room air by good work practices and use of laboratory hoods. When these are not sufficient, suitable respiratory protection should be provided. It is the responsibility of the laboratory supervisor to determine when such protection is needed, provide the proper protection, train laboratory workers in its use, and ensure that it is used properly.

Types of respirators

The term "respirator" refers to any device that covers the nose and mouth to prevent the inhalation of a harmful substance. All respirators are not equal! Respirators range from a disposable dust respirator to a

Table 2.2. Guide for selection of respirators

Type of hazard	Type of respirator
Oxygen deficiency	Self-contained breathing apparatus Hose mask with blower Combination air line respirator and auxiliary self-contained air supply or air storage receiver with alarm
Gas and vapour contaminants (Immediately dangerous to life or health)	Self-contained breathing apparatus Hose mask with blower Air-purifying full facepiece respirator with chemical canister (gas mask) Self rescue mouthpiece respirator (for escape only) Combination air line respirator and auxiliary self-contained air supply or air storage receiver with alarm
Gas and vapour contaminants (Not immediately dangerous to life or health)	Air line respirator Hose mask with blower Air-purifying half mask or mouthpiece respirator with chemical cartridge
Particulate contaminants (Immediately dangerous to life or health)	Self-contained breathing apparatus Hose mask with blower Air-purifying full facepiece respirator with appropriate filter Self rescue mouthpiece respirator (for escape only) Combination air line respirator and auxiliary self-contained air supply or air storage receiver with alarm
Particulate contaminants (Not immediately dangerous to life or health)	Air-purifying half mask or mouthpiece respirator with filter pad or cartridge Air line respirator Air line abrasive blasting respirator Hose mask with blower
Combination of gas, vapour and particulate contaminants (Immediately dangerous to life or health)	Self-contained breathing apparatus Hose mask with blower Air-purifying full facepiece respirator with chemical canister and appropriate filter (gas mask with filter) Self rescue mouthpiece respirator (for escape only) Combination of air line respirator and auxiliary self-contained air supply or air storage receiver with alarm
Combination of gas, vapour particulate contaminants (Not immediately dangerous to life or health)	Air line respirator Air line respirator Hose mask without blower Air-purifying half mask or mouthpiece respirator with chemical cartridge and appropriate filter

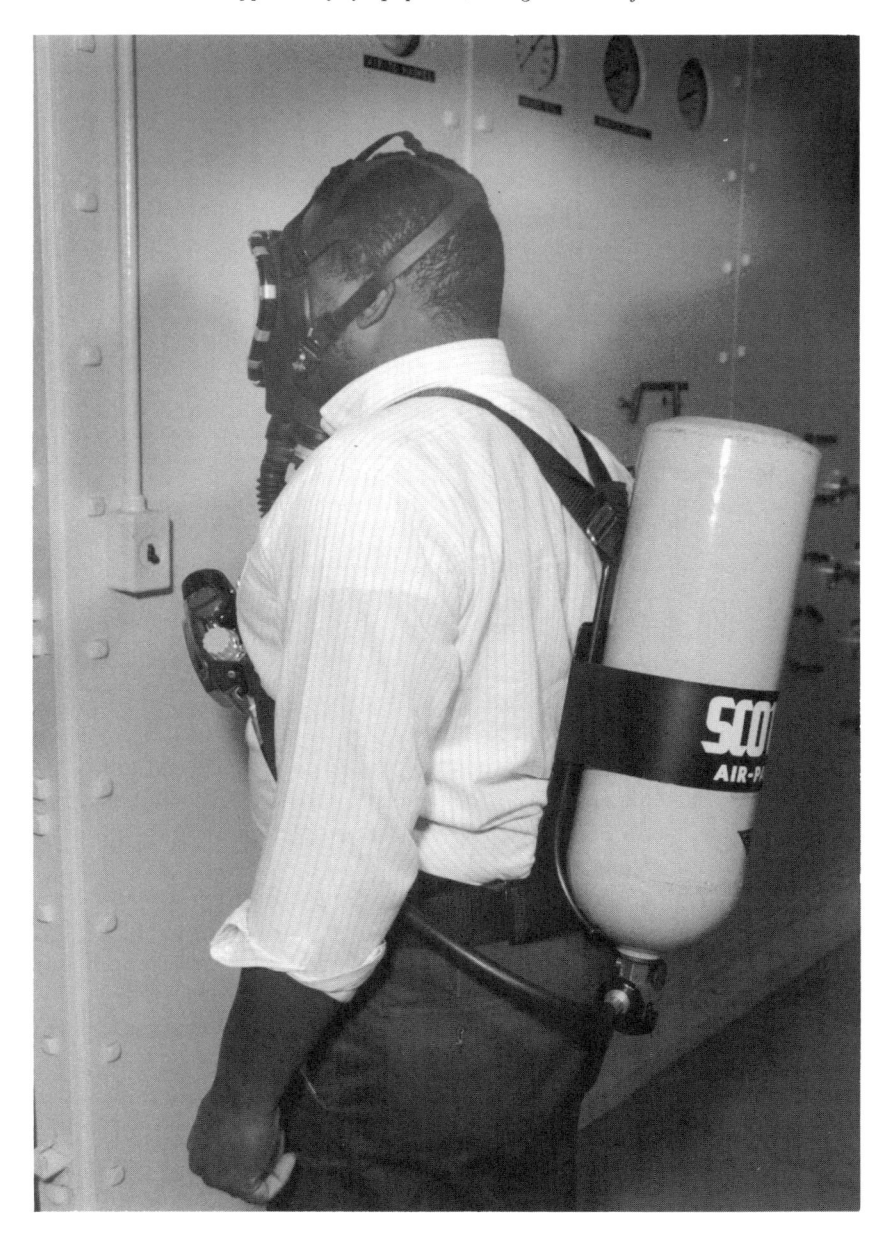

Fig. 4 Self-contained breathing apparatus (SCBA), with a full face mask connected to a cylinder of compressed air.

self-contained breathing apparatus. A guide to respirator selection is given in Table 2.2.

Self-contained breathing apparatus (SCBA). The first consideration in respirator selection is whether the intended use is for the "Immediately Dangerous to Life or Health" category. If this is the intended use, the only acceptable respirator is the SCBA (Figure 4). Although it is bulky and heavy, it is the only type of respiratory equipment suitable for emergency or rescue work. The equipment, which consists of a full face mask connected to a cylinder of compressed air, has no limitations to its use in areas of toxic contamination or oxygen deficiency. However, the air supply is limited to the capacity of the cylinder, which normally permits 5-30 minutes of use. Additional protective apparel may be required, depending on the nature of the hazard. SCBAs of the "pressure/demand" type, which always provide positive pressure within the mask (like supplied-air respirators, see below), are much preferred to the "demand type" for safety reasons.

Organizations with laboratories that use dangerous chemicals may require equipment of this type to be available for emergencies. If so, the laboratory should "fit-test" potential users and provide training in use of the equipment at least yearly.

Supplied-air respirators. These provide effective protection against a wide range of air contaminants (gases, vapours, and particulates) and can be used in oxygen-deficient atmospheres. Because failure of the air supply would expose the user to oxygen deficiency or another situation dangerous to life and health, an SCBA escape-cylinder assembly must be available for immediate use.

Supplied-air respirators bring fresh air through a length of hose to the facepiece of the respirator at a pressure high enough to cause the pressure inside the mask to be slightly greater than outside. Since the supplied air flows outward from the mask, contaminated air from the environment cannot readily enter the mask. However, it is possible for the mask to leak inward if the air supply is insufficient or if there are substantial leaks in the face-to-facepiece seal. Accordingly, fit testing is required before use.

The air supply for this type of respirator can be provided either by a compressed breathing air cylinder or an air compressor. Cylinder air should be certified breathing air. Air from a compressor must be kept free of contaminants such as particulates, oil mist and carbon monoxide. Because an over-heated oil compressor may generate dangerous concentrations of carbon monoxide, an oil-less compressor should be used. If this is not possible, an over-temperature alarm and a carbon monoxide monitor should be installed. Air quality must be checked periodically.

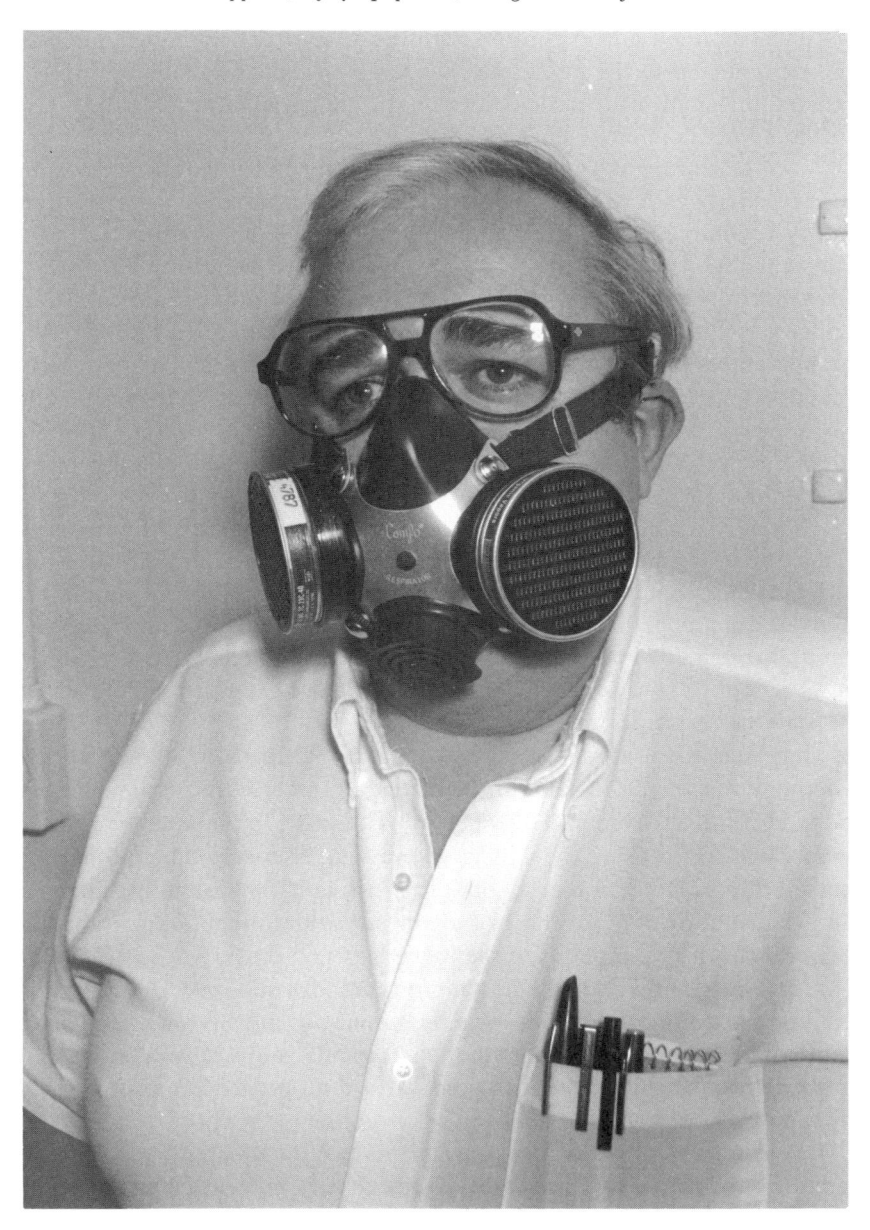

Fig. 5   Half-face respirator with replaceable cartridges selected to have good absorptive properties for the chemical vapour of concern.

Since supplied-air respirators require the user to drag long lengths of hose connected to the air supply, the range of their use is limited to the maximum length of hose specified by the manufacturer. They are also rather cumbersome to use.

Cartridge respirators. These are either half- or full-face respirators, usually with replaceable cartridges. Full-face respirators are normally recommended where the contaminant is an eye irritant or a substance which can be absorbed through the eye or surrounding membranes. Half-face respirators (Figure 5) suffice for other chemicals. Cartridges are specific for the gases, vapours, or particulates designated by the manufacturer and should not be used in concentrations above the recommended limit stated for a particular cartridge. Since these respirators do not supply air or oxygen, they cannot be used if the oxygen content of the air is less than 19.5%, or in atmospheres dangerous to life or health, or for rescue or emergency work.

Cartridge respirators work by trapping the contaminant(s). For particulates, breathing becomes difficult when the filter is saturated. A gas or vapour contaminant is collected on a sorbant material, for example, on activated charcoal for organic vapours. Because it is possible for significant breakthrough of contaminant to occur at a fraction of the cartridge capacity, it is important to know the potential workplace exposure and the length of time the respirator may be worn.

Cartridge respirators should be used only when the contaminant has adequate warning properties or the cartridge's service life has been determined for the environment of exposure. The cartridge should be replaced after each use to ensure the maximum available exposure time for the next period of use. Detection of an odour, or difficulty in breathing, indicates an exhausted or plugged filter or a concentration of contaminant higher than the absorbing capacity of the cartridge.

Care must be taken to assure that the cartridge will indeed remove the contaminant. For example, an organic vapour cartridge is not effective for removing methanol and some other compounds of low relative molecular mass. Particulate-removing cartridges afford no protection against gases or vapours.

These respirators must fit snugly to the face if they are to be effective. Conditions that prevent a face-to-facepiece seal, such as facial hair, long sideburns, or frames of eyeglasses, may permit contaminated air to bypass the filter. Tests for proper fit of the respirator on each potential user should be conducted before use in a contaminated area. Users must be trained in the use, care, and proper storage of the cartridge respirator, and in its use in situations of physical stress.

Disposable respirators. These are generally available with the same filter or sorbant as cartridge respirators. The difference is that the user is relieved of the responsibility for care and storage. The respirator is commonly used for a maximum period of 1 day and then discarded.

Procedures and training

Each area where respirators are likely to be used should have written information available that shows the limitations, fitting methods, and inspection and cleaning procedures for each type of respirator available. Personnel who may have occasion to use respirators in their work must be thoroughly trained in the fit-testing, use, limitations, and care of such equipment. Training should include demonstrations and practice in wearing, adjusting, and fitting.

Contact lenses should not be worn when a respirator is used, especially in a highly contaminated area. An employee should be examined by a physician for general fitness before beginning work in an area where a respirator may have to be worn.

Inspections

Respirators for routine use should be inspected before each period of use by the user, and periodically by the laboratory supervisor. Self-contained breathing apparatus should be inspected once a month and cleaned after each period of use. Defective units should not be used, but should be repaired by a qualified person or replaced promptly.

2.4.3. Safety showers and eyewash fountains

Safety showers

In areas where chemicals are handled safety showers should be provided for the purposes of immediate first aid treatment of chemical splashes and for extinguishing clothing fires. Every laboratory worker should learn where the safety showers are and how to use them, so that they can be found even with closed eyes. Safety showers should be tested regularly by laboratory personnel to ensure that they are operable. Shower heads which are not flushed periodically may build up deposits of scale, which can restrict the flow of water.

The shower should be capable of drenching the subject immediately and should be large enough to accommodate more than one person if necessary. It should have a quick opening valve requiring manual closing; a downward pull delta bar is satisfactory (Figures 6 and 7) if it is long enough, but chain pulls are not advisable because of the potential for persons to be hit in the face by them and the difficulty of grasping them in an emergency.

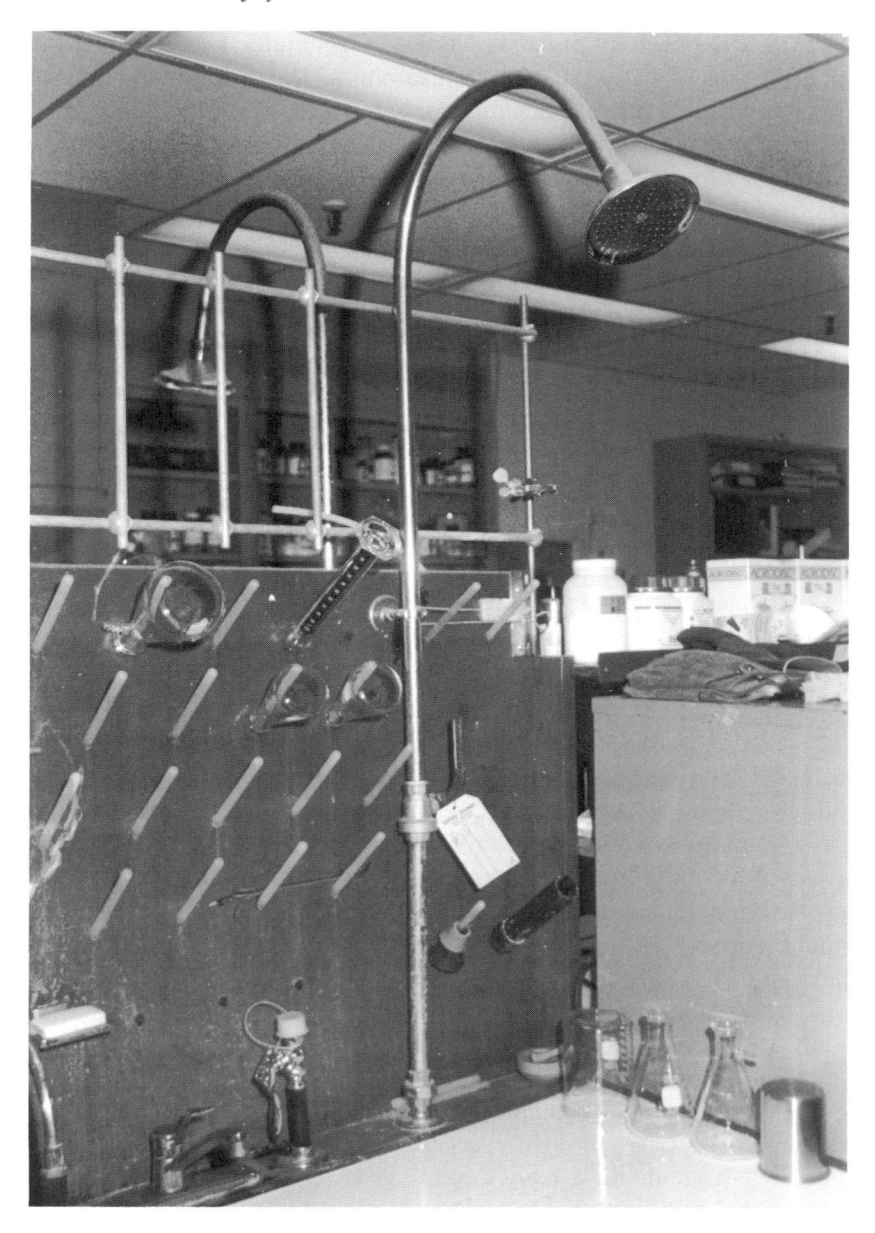

Fig. 6 Simple safety shower operated by a quick-opening valve; a label on the handle indicates the last date of inspection.

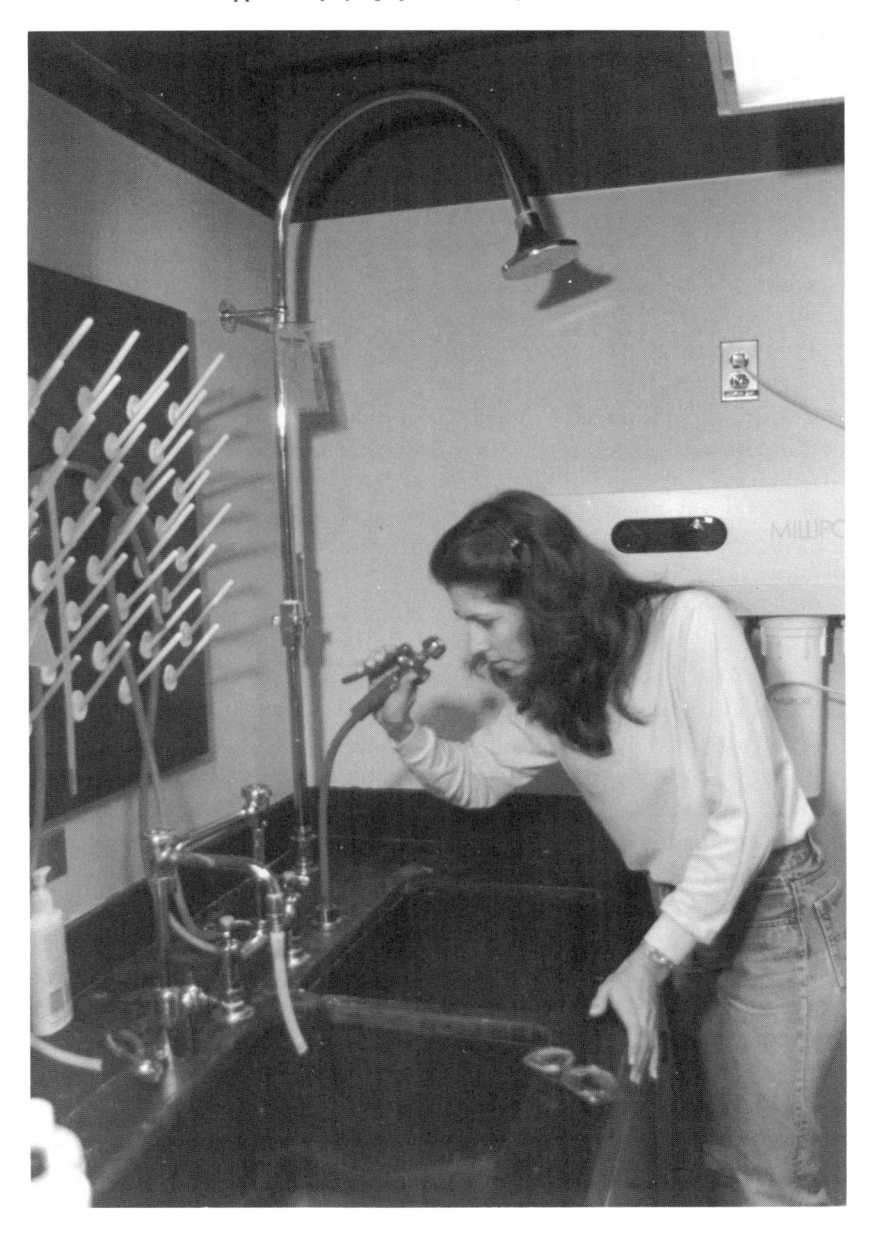

Fig. 7 Simple sink-mounted eyewash fountain operated by a quick-opening valve.

Eyewash fountains

Eyewash fountains (Figure 7) may be required if substances in use present a known or possible eye hazard. An eyewash fountain should provide a soft stream or spray of aerated water for about 15 minutes.

2.4.4. Other safety equipment

Safety shields

Safety shields should be used for protection against possible explosions or splash hazards. Laboratory equipment should be shielded on all sides so that there is no line-of-sight exposure of personnel.

The conventional laboratory exhaust hood (fume cupboard) constitutes a readily available "built-in" shield, provided that its opening is covered by closed doors. However, a portable shield may also be needed during manipulations, particularly with hoods that have vertical rising sashes rather than horizontal-sliding doors.

Portable shields can be used to protect against hazards of limited severity, e.g., small splashes, heat, and fires. A portable shield, however, provides no protection at the sides or back of the equipment.

Portable shields are not heavily weighted and may topple toward the worker when there is a blast, perhaps hitting the worker or permitting exposure to flying objects. A fixed shield that completely surrounds the experimental apparatus can afford protection against minor blast damage.

Polymethyl methacrylate, polycarbonate, polyvinyl chloride, and laminated safety plate glass are all satisfactory transparent shielding materials. Where combustion is possible, the shielding material should be nonflammable or slow burning. If it can withstand the working blast pressure, laminated safety plate glass may be the best material for such circumstances. When cost, transparency, high tensile strength, resistance to bending loads, impact strength, shatter resistance, and burning rate are considered, polymethyl methacrylate offers an excellent overall combination of shielding characteristics. Polycarbonate is much stronger and is self extinguishing after ignition, but it is readily attacked by organic solvents.

Stretchers

Although stretchers are sometimes provided in areas where chemicals are handled, untrained personnel should use them only in life-threatening situations. It is generally best not to move a seriously injured person until qualified medical help arrives.

2.4.5. Storage and inspection of emergency equipment

It is often useful to establish a central location for storage of emergency equipment apart from fire appliances. Such a location may contain the following items:

1. Self-contained breathing apparatus.

2. Safety belt with rope, to maintain contact physical with a rescuer entering a laboratory under emergency conditions.
3. Blankets for covering injured persons.
4. Stretchers.
5. First aid equipment for special situations such as exposure to cyanide, where immediate first aid is required.

Safety equipment should be inspected regularly (e.g., every 3 to 6 months) to ensure that it will function properly when needed.

Fire extinguishers should be inspected for broken seals, damage, and low pressure. The mounting of the extinguisher and its accessibility should also be checked. An inspection tag showing the date of the latest inspection and the name of the inspector should be attached to or near the equipment. Some types of extinguishers must be weighed annually, and periodic hydrostatic testing may be required.

Self-contained breathing apparatus should be checked at least once a month, and after each use, to determine whether proper air pressure is being maintained. The examiner should look for signs of deterioration or wear of rubber parts, harness, and hardware, and make certain that the apparatus is clean.

Safety showers and eyewash fountains should be tested regularly.

2.5. EMERGENCY PROCEDURES

The following emergency procedures are recommended in the event of fire, explosion, or other laboratory accident. These procedures are intended to limit injuries and minimize damage if an accident should occur.

1. Render assistance to persons involved and, if necessary, remove them from exposure to further injury.
2. Warn personnel in adjacent areas of any potential hazards to their safety.
3. Render immediate first aid; appropriate measures include washing under a safety shower, administering oxygen and artificial resuscitation if breathing has stopped, and the use of a cyanide first aid kit for cyanide exposure.
4. Extinguish a small fire with a portable extinguisher. Turn off nearby apparatus and remove combustible materials from the area. In case of a larger fire, contact the appropriate fire department promptly.

In case of medical emergency, laboratory personnel should remain calm and do only what is necessary to protect life:

1. Summon medical help immediately.

2. Do not move an injured person unless there is danger of further harm.
3. Keep any injured person warm. If possible designate one individual to remain with the injured person. The injured person should be within sight, sound, or physical contact of that person at all times.
4. If clothing is on fire, the person should be knocked to the floor and rolled around to smother the flames. Alternatively, if a safety shower is close, douse with water.
5. If a chemical has been spilled on the body, flood the exposed area with sufficient running water from the safety shower and immediately remove any contaminated clothing.
6. If a chemical has entered the eye, immediately wash the eyeball and the inner surface of the eyelid with plenty of water for 15 min, preferably using an eyewash fountain. *Forcibly hold the eye open to wash thoroughly behind the eyelids.*

2.5.1. Preparing for emergencies

It is the responsibility of every laboratory organization to establish a specific emergency plan. Such a plan should include description of evacuation routes and shelter areas, location of medical facilities, and procedures for reporting all accidents and emergencies. It should be reinforced by drills and simulated emergencies.

Evacuation procedures

Evacuation procedures should be established and communicated to all personnel.

1. *Emergency alarm system* - There should be a system to alert personnel of an emergency that may require evacuation of an area or building. Laboratory personnel should be familiar with the location and operation of this equipment. A system should be established to relay telephone alert messages. The names and telephone numbers of personnel responsible for each laboratory or other area should be prominently posted in case of an emergency outside regular working hours.

   Isolation areas (e.g., cold, warm, or sterile rooms) should be equipped with alarm or telephone systems that can be used to alert outsiders to the presence of a worker trapped inside, or to warn workers inside of the existence of an emergency outside that requires evacuation. Where an unusually toxic substance is handled, it may be desirable to have a monitoring and alarm system so that, if the concentration of the substance

in the work environment exceeds a set limit, an alarm is sounded to warn the laboratory workers to evacuate the area.
2. *Evacuation* - Evacuation routes should be established and communicated to all personnel. An outside assembly area for evacuated personnel should be designated, with plans for taking roll call to ascertain that all personnel are present.
3. *Shutdown procedures* - Brief guidelines for shutting down operations during an emergency or evacuation should be communicated to all personnel.
4. *Return and start-up procedures* - Return procedures to ensure that personnel do not return to the laboratory until the emergency is ended, and start-up procedures that may be required for some operations, should be prominently displayed and reviewed regularly.
5. *Drills* - All aspects of the emergency procedure should be tested regularly (e.g., every 6 or 9 months), and trials of evacuations should be held periodically.

Medical facilities

Laboratories that do not have a full time medical staff should have personnel trained in first aid available during regular working hours to render assistance until medical help can be obtained. These people should know where to seek medical help, and they should be conversant with the appropriate emergency procedures.

An emergency room staffed with medical personnel specifically trained in proper treatment of chemical exposures should be readily accessible. For small laboratories, prior arrangement with a nearby hospital or emergency room may be necessary to ensure that treatment will be available promptly. Proper and speedy transportation of the injured to the medical treatment facility should be available. Emergencies that should be anticipated include the following:

1. Thermal and chemical burns.
2. Cuts and puncture wounds from glass or metal, including possible chemical contamination.
3. Skin irritation by chemicals.
4. Injuries to the eyes from splashed chemicals or solids.
5. Poisoning by ingestion, inhalation, or skin absorption.
6. Asphyxiation (chemical or electrical).

Accident and emergency reporting

A system should be established to ensure that accidents or emergencies are reported promptly to the persons responsible for safety matters. Such reports, sometimes required by law, help uncover hazards that can be corrected. The report should be in a written form and a copy retained as part of the safety record programme.

2.5.2. Fires and explosions

Small fires that can be extinguished easily without evacuating the building or calling the fire department are among the most common laboratory incidents. However, even a minor fire can quickly become a serious problem. The first few minutes after discovery of a fire are critical in preventing a larger emergency. The following actions should be taken by laboratory personnel in case of a minor fire:

1. Alert other personnel in the laboratory and send someone for assistance.
2. Attack the fire immediately, but never attempt to fight a major fire alone. A fire in a small vessel can often be suffocated by covering the vessel with an inverted beaker or a clock glass or by the judicious use of a fire blanket. Alternatively, direct the discharge of an appropriate extinguisher (Section 2.4.1) at the base of the flame.
3. Avoid being trapped by fire; always fight a fire from a position accessible to an exit.

If there is any doubt whether the fire can be controlled by locally available personnel and equipment, the following actions should be taken:

1. Notify the fire department and activate the emergency alarm system.
2. Confine the emergency (close hood sashes, doors between laboratories, and fire doors) to prevent further spread of the fire.
3. Assist injured personnel (provide first aid or transportation to medical aid if necessary as outlined in Section 2.6).
4. Evacuate the building to avoid further danger to personnel.

In case of an explosion, immediately turn off burners and other heating devices, stop any operations in progress, assist in treating persons, and vacate the area until it has been decontaminated.

It is the responsibility of the laboratory supervisor to determine whether an unusual hazard exists that requires more stringent safety precautions. In large laboratories, or where risk is high, designated fire-fighting teams may be necessary to minimize risk. Special arrangements with local fire departments to warn them of the hazards of chemical fires may be desirable.

2.5.3. Chemical spills

Where a large-scale spillage may occur, such as in a storeroom or similar work area, an emergency procedure should be prepared for containing a spilled chemical with minimal damage. A spill control policy should include consideration of the following points:

1. *Prevention* - storage, operating procedures, monitoring, inspection, and personnel training.
2. *Containment* - engineering controls on storage facilities and equipment.
3. *Clean-up* - training of designated personnel in counter measures to help reduce the impact of a chemical spill.
4. *Reporting* - provisions for reporting a spill both internally (to identify a controllable hazard) and externally (for example, to a regulatory agency).

For dealing with small-scale laboratory spillage, there should be supplies and equipment in the laboratories or nearby. The clean-up supplies should include neutralizing agents such as sodium carbonate and sodium hydrogen sulfate and absorbents such as vermiculite and sand. A 1:1:1 (w/w/w) mixture of soda ash or calcium carbonate, cat litter (bentonite clay), and sand is a useful absorbent for spillage of most acidic or neutral materials. Paper towels and sponges may also be used as absorbent clean-up aids, although this should be done cautiously. For example, paper towels used to clean-up a spilled oxidizing agent may later ignite. Appropriate gloves (Section 2.2) should be worn when wiping up a highly toxic material with paper towels. When a spilled flammable solvent is absorbed in vermiculite or sand, the resultant solid is highly flammable and gives off flammable vapour; this solid must therefore be properly contained in a covered metal pail or removed to a safe place (temporarily, to a hood).

Commercial spillage kits are available that have instructions, absorbents, reactants, and protective equipment. These should be located strategically around work areas near the fire extinguishers.

The strategy to be followed using a spill kit is as follows:

1. Warn all personnel in the immediate vicinity.
2. Minimize further spillage and contain the affected area if it is safe to do so.
3. Evacuate all non-essential personnel from the spillage area.
4. If the spilled material is flammable, turn off ignition and heat sources.
5. Avoid breathing vapours of the spilled material; if necessary, use a respirator (Section 2.4.2).
6. Leave on, or establish, exhaust ventilation if it is safe to do so.
7. Secure supplies to effect clean-up.
8. During clean-up, wear appropriate apparel (Sections 2.2, 2.3).
9. Notify the safety officer or other responsible person.

46   *Chemical Safety Matters*

Handling of leaking gas cylinders is discussed in Section 10.5. Handling of spilled liquids or solids is discussed below.

Handling of spilled liquids
1. Confine the spillage to a small area. Do not allow it to spread.
2. For small quantities of inorganic acids or bases, use a neutralizing agent or an absorbent mixture (e.g., soda ash with diatomaceous earth). For small quantities of other materials, absorb the spillage with a non-reactive material such as vermiculite, dry sand, or towels. Scoop the mixture into a pail and place it in a hood.
3. For larger amounts of inorganic acids and bases, flush with large amounts of water, provided that the water will not cause additional damage or chemical reaction. Flooding is not recommended in storerooms or in areas where water-reactive chemicals are present.
4. Mop up the spillage, wringing out the mop in a sink or in a pail equipped with rollers.
5. Carefully pick up and clean any cartons or bottles that have been splashed or immersed.
6. If necessary, clean the area with a suction cleaner approved for the material involved, remembering that the exhaust of a vacuum cleaner can create aerosols and thus should be vented to a hood or through a filter.
7. If the spilled material is extremely volatile, let it evaporate and be exhausted by the mechanical ventilation system, provided that the associated electromechanical system are spark-free.
8. Dispose of residues by safe procedures (Chapter 10).

Handling of spilled solids
Generally, sweep spilled solids of low toxicity into a dust pan and place them in a solid-waste container for disposal. Additional precautions, such as the use of a suction cleaner equipped with a high efficiency particulate air filter, may be necessary when cleaning up spills of more highly toxic solids.

2.5.4. Other emergencies

Laboratories should be prepared for hazards resulting from severe weather or loss of a utility service. Loss of the water supply, for example, can render safety showers, eyewash fountains, and sprinkler systems inoperative. All hazardous laboratory work should cease until the service is restored.

2.6. FIRST AID

First aid is the immediate care of a person who has been injured or has suddenly become ill. It is intended to prevent further illness, injury or death and to relieve pain until medical aid can be obtained. The objectives of first aid are to:

1. Control conditions that might endanger life.
2. Prevent further injury or chemical contamination.
3. Relieve pain and treat for shock.
4. Make the patient as comfortable as possible.

The initial responsibility for first aid rests with the first persons at the scene, who should react quickly but in a calm and reassuring manner. The person assuming responsibility should immediately summon medical help, being explicit in reporting suspected types of injury or illness and in requesting assistance. An injured person should not be moved except under medical direction or when necessary to prevent further injury. Laboratory workers should be encouraged to obtain training in first aid and cardiopulmonary resuscitation.

2.6.1. Pulmonary resuscitation

If the subject is unresponsive and no breathing movements are apparent, begin mouth-to-mouth resuscitation immediately. Delay increases the risk of serious disability or death.

1. Place the patient flat on his or her back on the floor and kneel at the side.
2. Establish an airway. Check the patient's mouth with a finger to be sure that no obstruction is present and then tip the patient's head back until the chin points straight up.
3. Pinch the patient's nostrils, and begin mouth-to-mouth resuscitation by taking a deep breath and placing the mouth over the patient's mouth so as to make a leakproof seal. Blow breath into the patient's mouth until it is seen that the chest rises.
4. Remove the mouth and allow the patient to exhale.
5. Repeat the procedure at a rate of once every 5 seconds.

2.6.2. Cardiac resuscitation

In the unresponsive patient, check for a cardiac pulse, locate the larynx (Adam's apple) with the tips of the fingers and slide them into the groove between it and the muscle at the side of the neck. If no pulse is felt, circulation must be re-established within 4 minutes to prevent brain damage.

1. With the patient flat on the back, kneel at the waist, facing the head.
2. Place the heel of one hand over the heel of the other hand on top of the patient's breastbone about 3 cm above its lower tip.
3. Shift body weight onto the patient's chest so as to compress it at least 4 cm, then remove the pressure.
4. Continue at a rate of 80 times/minute.

2.6.3. Heavy bleeding

Heavy bleeding is caused by injury to one or more large blood vessels. Lay the patient down. Control bleeding by applying firm pressure directly over the wound with a clean handkerchief, cloth, or the hand. A tourniquet should be applied only in cases of an amputation or other injury to a limb in which there is no other way to stop the bleeding. If a tourniquet is used, a record of the time it was applied should be kept as information for medical personnel when they arrive.

2.6.4. Shock

Shock usually accompanies severe injury. The signs of shock include pallor, a cold and clammy skin and beads of perspiration on the forehead and palms or hands, together with weakness, nausea or vomiting, shallow breathing, and a rapid pulse that may be too faint to be felt at the wrist. The following procedure for the treatment of shock should be followed:

1. Correct the cause if possible, for example by controlling bleeding.
2. Keep the patient lying down. If there are no contra-indications (e.g., a head injury), elevate the patient's legs.
3. Keep the patient's airway open. If the patient is about to vomit, turn the head to the side.
4. Keep the patient warm.

2.6.5. Miscellaneous symptoms, illnesses and injuries

After requesting medical aid, the following points should be addressed in specific emergencies:

1. *Abdominal pain* - Keep the patient quiet. Give nothing by mouth.
2. *Back and neck injuries* - Keep the patient absolutely quiet. Do not move or lift the head unless absolutely necessary.
3. *Chest pain* - Keep the patient calm and quiet. Place in the most comfortable position (usually half-sitting).
4. *Convulsion or epileptic seizure* - Place the patient on the floor or a couch. Do not restrain the patient's movements except to

Protective apparel, safety equipment, emergencies and first aid 49

   prevent injury. Do not place a blunt object between the teeth, or put any liquid in the mouth, or slap the patient or douse with water.
5. *Electric shock* - Throw the switch to turn off the current. Do not touch the victim or clothing until contact with the current source has been eliminated. Begin mouth-to-mouth resuscitation if respiration has ceased.
6. *Fainting* - Simple fainting can usually be ended quickly by laying the person down.
7. *Unexplained unconsciousness* - Look for emergency medical identification around the person's neck or wrist or in the wallet. Keep the person warm, lying down, and quiet until consciousness returns. Do not move the head if there is bleeding from the nose, mouth, ear, or eyes. Do not give anything by mouth. Keep the person's airway open to aid breathing. Do not cramp the neck with a pillow.

2.7. CHEMICAL INGESTION OR CONTAMINATION
Ingestion of chemical

Encourage the person to drink large amounts of water while awaiting medical assistance. Attempt to learn exactly what substance was ingested and inform the medical staff as soon as possible or, if appropriate, the local poison control centre.

Chemical spilled on the body over a large area

Quickly remove all contaminated clothing while using the safety shower. Seconds count, and no time should be wasted because of modesty. Immediately flood the affected body area with cold water for at least 15 minutes. Resume if pain returns. Wash the chemical from the skin by using a mild detergent or preferably soap and water. Do not use a neutralizing chemical, unguent, or salve.

Chemical on the skin over a localized area

Immediately flush the area with cold water and wash with soap and water. If there is no visible burn, scrub with warm water and soap, after removing any jewellery near the affected area. If there is likely to be a delayed action, obtain medical attention promptly and explain carefully what chemical was involved. Note that the physiological effects of some chemicals, such as ethyl bromide, may be delayed 48 hours or more.

3

Laboratory Ventilation

Laboratories that use chemicals vary from spacious well designed facilities to those that consist of a single room that has been designated as a laboratory and has little or no provision for ventilation. The needs for ventilation of these different types of laboratories will vary from the provision of simple comfort for the occupants to the control of highly toxic volatile substances. This discussion will centre on ventilation for the control of toxic chemicals. However, the overall performance of laboratory workers will also benefit from ventilation systems that control the temperature, humidity, and concentration of odorous materials in the laboratory. Specialized studies of laboratory hoods for local exhaust ventilation are included in the *Bibliography*.

3.1. GENERAL LABORATORY VENTILATION
General ventilation is concerned with the quantity and quality of the air supplied to the laboratory. The overall ventilation system should ensure that the laboratory air is continually being replaced so that concentrations of odorous or toxic substances do not increase during the working day. A ventilation system that changes the room air 4-12 times per hour is normally adequate, provided that auxiliary local exhaust systems (Sections 3.2, 3.3) are available and are used as the primary method for controlling concentrations of airborne substances.

In all cases, the movement of air in the general ventilation system for a building should be from the offices, corridors, and conference rooms into the associated laboratories. All air from laboratories should be exhausted outside and not recycled. Thus, the air pressure in the laboratories should always be negative with respect to the rest of the building. The air intakes for a laboratory building should be in a location that reduces the possibility that the input air will be contaminated by the exhaust air from either the same building or any other nearby laboratory building. One common arrangement is to locate all the laboratory hood exhaust vents (the usual exhaust ports for laboratories) on the roof of the laboratory building and for the air intake port to be placed on a side of the building where local air movement is unlikely to mix exhaust and intake air.

3.1.1. Evaluation of general laboratory ventilation

Each laboratory should be evaluated for the quality and quantity of general ventilation. This evaluation should be repeated periodically, and at any time when a change is made in the general building ventilation system or in some aspect of the local laboratory ventilation. This evaluation should begin by visualizing the pattern of air movement entering and within the laboratory.

Air flow paths into and within a room can be determined by observing smoke patterns. Convenient smoke sources are commercial smoke tubes available from local safety and laboratory suppliers. If the general laboratory ventilation is satisfactory, the movement of air from the corridors to the hoods or other exhaust ports should be relatively uniform. There should be no areas where the air remains static or has unusually high airflow velocities. If areas that have little or no air movement are found, a ventilation engineer should be consulted and appropriate changes made in input or output ports to correct the deficiencies. Alternatively, warning signs of inadequate ventilation should be posted in such areas.

At low concentrations, most chemical vapours and gases tend to rise with warm air currents and become diluted with the general room air. Air movement in large laboratories is normally multidirectional and typically has a velocity of about 0.1 m/sec. This diverse air movement is a result of the movement of people and the effects of air intakes, air exhausts, and eddy currents around benches and other fixed objects. The net result is that the general air composition is rather uniform. This mixing of room air does not mean that isolated static air spaces cannot be found, but rather that the general area occupied by the laboratory workers tends to be uniform in composition unless there are serious defects in the locations of input and output ports.

It is important to realize that, up to the capacity of the exhaust system, the rate at which air is exhausted from the laboratory will equal the rate at which input air is introduced. Thus, decreasing the flow rate of input air (perhaps to conserve energy) will decrease the number of air changes per hour in the laboratory, the face velocities of the hoods, and the capture velocities of all other local ventilation systems.

The measurement of airflow rates requires special instruments and personnel trained to use them. Pitot tubes are used for measuring duct velocities. Anemometers or velometers are used to measure airflow rates within rooms and at the faces of input or exhaust ports. These instruments are available from safety supply companies or laboratory suppliers. The proper calibration and use of these instruments and the evaluation of the data are a separate discipline. Consultation with an industrial hygienist or a ventilation engineer is recommended whenever a serious ventilation problem is suspected or when a decision must be made on changes in the ventilation system to achieve a proper balance of input and exhaust air.

If a reasonable measure of the exposure to a specific substance is required, it can be obtained by fitting a laboratory worker with a portable air-sampling device. Analysis of the uptake of the sampling device provides a measure of the average concentration of a specific substance to which the worker was exposed. Suitable air-sampling devices are available from various safety supply companies and laboratory suppliers.

3.1.2. Use of general laboratory ventilation

As mentioned above, general laboratory ventilation is intended primarily to increase the comfort of laboratory workers and to provide a supply of air that will be exhausted by a variety of auxiliary local ventilation devices such as hoods, vented canopies, and vented storage cabinets. This ventilation provides only modest protection from toxic gases, vapours, aerosols, and dusts, especially if they are released into the laboratory in a significant quantity. The cardinal rules for safety in working with toxic substances are that all work should be performed in such a way that no material comes in contact with the skin and that the quantity of vapour or dust that might produce adverse toxic effects is prevented from entering the general laboratory atmosphere. Thus, operations such as chemical reactions, heating or evaporating solvents, and the transfer of chemicals from one container to another should normally be performed in a hood. If especially toxic or corrosive vapours are evolved, these should be passed through scrubbers or adsorption vessels. Toxic substances should be stored in cabinets fitted with auxiliary local ventilation. Laboratory apparatus that may discharge toxic vapours (e.g., vacuum pump exhausts, gas chromatograph exit ports, liquid chromatographs, distillation columns) should be vented to an auxiliary local exhaust system such as a canopy. Samples placed in instruments or apparatus where auxiliary local ventilation is not practical (e.g., balances, spectrometers, refrigerators) should be kept in closed containers. In other words, laboratory workers should regard the general laboratory atmosphere only as a source of air to breathe and as a source of input air for auxiliary local ventilation systems.

3.2. USE OF LABORATORY HOODS

Although many laboratory workers regard hoods solely as local ventilation devices to be used to prevent toxic, offensive, or flammable vapours from entering the general laboratory atmosphere, hoods offer two other significant types of protection. Placing a chemical reaction system within a hood, especially with the hood sash closed, also places a physical

Laboratory ventilation

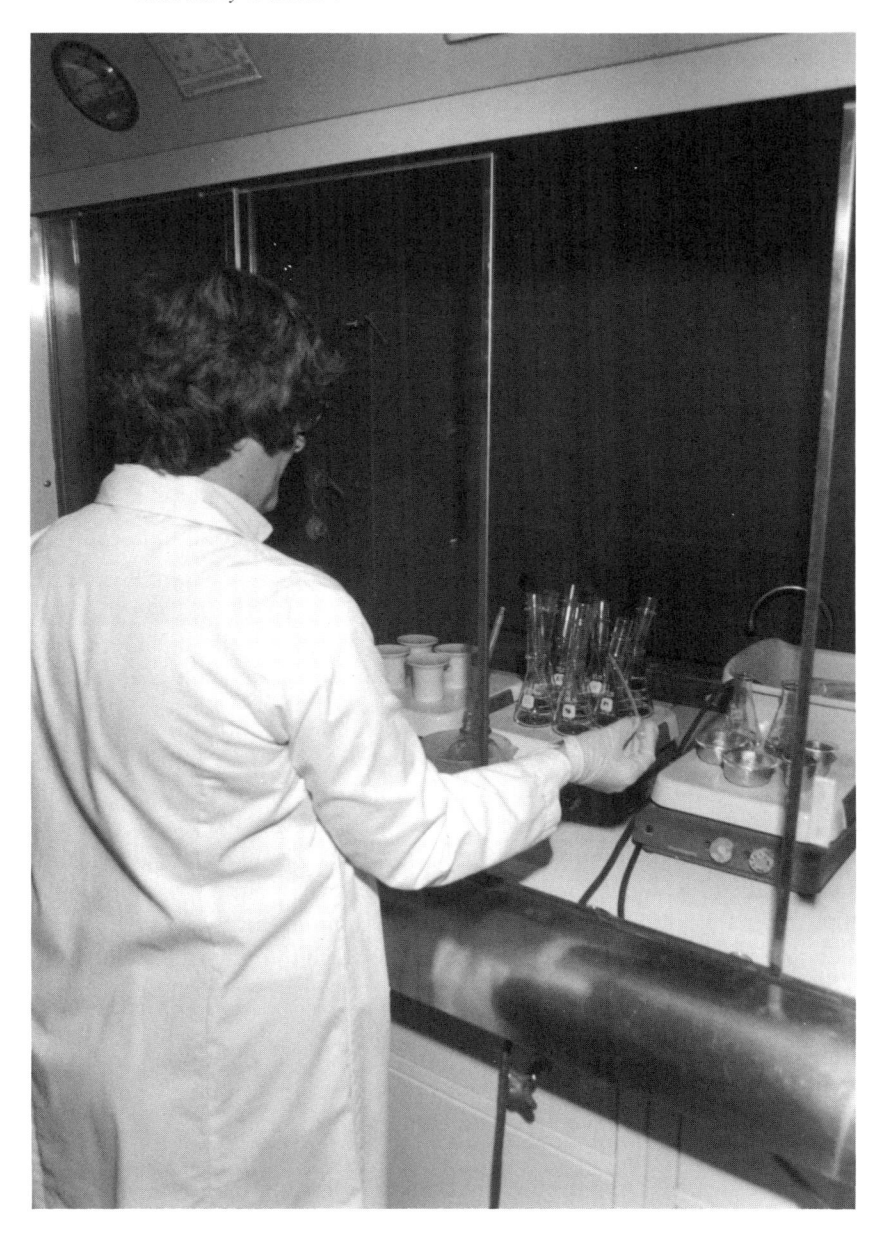

Fig. 8 Good practice in working with a hazardous reaction mixture. The flasks are set back in the hood (fume cupboard) and the chemist is protected by the sliding sash.

Fig. 9   Hood damaged by a small explosion and fire, showing the effectiveness of the safety screens.

barrier between the worker and the chemical system (Figure 8). This barrier can afford the laboratory workers significant protection from hazards such as chemical splashes or sprays, fires, and explosions (Figure 9). Furthermore, the hood can provide an effective containment device for accidental spills of chemicals. In a laboratory where workers spend most of their time working with chemicals, there should be at least one hood for each two workers, and the hoods should be large enough to provide each worker with at least 0.75 m of working space across the face. The optimum arrangement is to provide each laboratory worker with a separate hood. In circumstances where this amount of hood space cannot be provided, there should be reasonable provisions for other types of local ventilation, and special care should be exercised in monitoring and restricting the use of hazardous substances.

The following factors should be taken into account in the use of hoods:

1. Hoods should be considered as back-up safety devices that can contain and exhaust toxic, offensive, or flammable materials if the design of an experiment fails so as to cause vapours or dusts to escape from the apparatus being used. Hoods should not be regarded as a means of disposing of chemicals. Thus, apparatus used in hoods should be fitted with condensers,

traps, or scrubbers to contain and collect waste solvents, toxic vapours, or dusts. Highly toxic or offensive vapours should be scrubbed or absorbed before the exit gases are released into the hood exhaust system.

2. Hoods should be evaluated before use to ensure adequate face velocities, typically 0.3-0.5 m/s. Excessive turbulence should be shown to be absent (Section 3.2.3). Further, some continuous monitoring device for adequate hood performance should be present and should be checked before each hood is used.

3. Except when adjustments to apparatus within the hood are being made, the hood should be kept closed. Vertical sashes should be down, and horizontal ones closed. Sliding doors should not be removed from horizontal-sliding door hoods. Keeping the face opening of the hood small improves the overall performance of the hood.

4. The airflow pattern, and thus the performance of a hood, depends on such factors as placement of equipment in the hood, room draughts from open doors or windows, persons walking by, and even the presence of the user in front of the hood. For example, moving a large piece of apparatus 5-10 cm back from the front edge into the hood can reduce the vapour concentration at the face by 90%.

5. Hoods are not intended primarily as storage areas for chemicals. Materials stored in them should be kept to a minimum. Stored chemicals should not block vents or alter airflow patterns. Whenever practical, chemicals should be moved from hoods to vented cabinets for storage.

6. Light objects and materials (such as paper) should not be permitted to enter the exhaust ducts because they can lodge in the ducts or fans and adversely affect their operation.

7. An emergency plan should be available in the event of ventilation failure (e.g., power failure), or any other unintended occurrence such as fire or explosion.

8. If laboratory workers are certain that adequate general laboratory ventilation will be maintained when the hoods are not running, then hoods not in use should be turned off to conserve energy. If any doubt exists, however, or if toxic substances are being stored in the hood, it should be left on. Energy can also be conserved by the use of variable volume hoods that modulate exhaust flow according to sash or door position.

9. The fact that the hood fan motor can be heard to be running is no guarantee that the hood is operating properly (there may

be a fault in the fan drive). A pressure device should be installed in the exhaust which triggers an alarm if the correct exhaust pressure is not maintained.
10. The hood exhaust should pass through a charcoal filter when volatile carcinogens are being used.
11. Hood exhaust filters should be fitted where tritium, ^{125}I or ^{131}I are being used.

3.2.1. General requirements and evaluation of laboratory hoods

The hood is the best-known local exhaust device used in laboratories. It is, however, only one part - albeit the principal exhaust port - of the total ventilation system and should not be considered as separate from the total system. Its performance will be strongly influenced by other features in the general ventilation system. Studies of the protection that hoods afford to laboratory workers have shown the importance of factors such as the volume of input air to the laboratory, the location of the laboratory air input ports, the location of the hood within the laboratory, and the placement of apparatus within the hood. Any efforts to correct poor hood performance should involve consideration of all these factors.

3.2.2. Hood design and construction

Laboratory hoods and the associated exhaust ducts should be constructed of non-flammable materials. They should be equipped with either vertical sashes or horizontal doors that can be closed. Welded steel construction is recommended for the sash frame. The glazing should be laminated safety-glass (at least 5.5 mm thick) or other equally safe material that will not shatter if there is an explosion within the hood. Gas and water control valves, electrical points, and similar fixtures should be located outside the hood to minimize the need to reach inside.

In recent years, the "supplementary air hood" has become popular. This device directs a blanket of air vertical to the hood face between the operator and the sash. As much as 70% of the total air exhausted by the hood can be taken from a supplementary unheated or uncooled air source, resulting in considerable savings in energy. However, careful balancing of both the velocity and direction of the incoming air is required to achieve an even air distribution across the hood face. Consequently, the design and installation of such a system should not be attempted without the aid of a qualified ventilation engineer.

Although hoods are most commonly considered as devices for controlling concentrations of toxic vapours, they can also serve to dilute and exhaust flammable vapours. However, a hood can be used effectively or it can be overloaded and misused, resulting in spillage of vapour into the general room air. An overloaded hood may contain an explosive mixture of air and a flammable vapour (Section 7.1). Both the hood designer and the user should recognize this hazard and eliminate possible sources of ignition within the hood and its ducting.

In some situations, the materials that might be exhausted by a hood are sufficiently toxic that they should not be expelled into the outside air. Whenever possible, experiments involving such materials should be designed so that the toxic materials are collected in traps or scrubbers rather than being released into the hood.

The ducts used for hood exhaust air should be used specifically for that purpose and not combined with other ventilation ducts within the building. It is best to have separate ducts for each hood in order to eliminate the possibility of toxic vapours exhausted from one hood being channelled into an unused hood and re-entering the general laboratory atmosphere. If several hoods are connected to a common exhaust duct, then some fail-safe arrangement should be provided to ensure that all of them are continually exhausting air when any one of them is in use.

3.2.3. Evaluation of hood performance

All hoods should be evaluated for performance when they are installed. The design specifications for uniform airflow across the hood face as well as for the total exhaust air volume should be achievable. Equally important is the evaluation of operator exposure. This should be repeated every time there is a change in the ventilation system. Thus, changes in the total volume of input air, changes in the locations of air input ports, or the addition of other auxiliary local ventilation devices (e.g., more hoods or vented cabinets) require a re-evaluation of the performance of all hoods in the laboratory.

The first step in the evaluation of hood performance is the use of a smoke tube or similar device to determine that the hood is actually exhausting air. The second step is to measure the velocity of the airflow at the face of the hood, and the third step is to determine the uniformity of air delivery to the hood face by making a series of face velocity measurements taken in a grid pattern. Sets of measurements should be made with the hood sash fully opened and in one or more partially closed positions.

The total volume of air being exhausted by a hood is the product of the average face velocity and the area of the hood face opening. If the hood and general ventilating system are properly designed, face velocities in the range of 0.3-0.5 m/s will provide a laminar flow of air over the floor and sides of the hood. Higher face velocities, 0.6 m/s or more, which exhaust the general laboratory air at a greater rate, are both wasteful of energy and likely to degrade hood performance by creating air turbulence at the hood face and within the hood. Such air turbulence can cause vapours within the hood to spill out into the laboratory atmosphere.

The second aspect of hood performance that should be evaluated is the presence or absence of air turbulence at the face and within the hood. The observation of smoke patterns is used for this purpose. Visible fumes or smoke can be generated by using commercial smoke tubes, aerosol generators, or cotton swabs dipped in titanium(IV) chloride. Each hood should be tested for air turbulence and capture effectiveness before it is used, and again after any change is made in the overall ventilation system of the laboratory. If there is excessive turbulence or if the hood fails to capture smoke, changes may be required in the hood face velocity, the location of the air input ports, the physical location of the hood, or the volume of input air.

The location of the hood within the laboratory will affect its performance. If it is placed so that cross-draughts from the movement of people or air currents from open windows or doors exceed the capture velocity, material may be drawn from the hood into the room. Often, air turbulence at the hood face is best diminished by relocating the air input ports or by adding external baffles near the hood face. A qualified ventilation engineer should be consulted for aid in solving such problems.

Another factor that influences the performance of a hood is the amount and location of equipment in it. As in the general laboratory, the air in the hood moves in all directions. Hot-plates, heating mantles, and equipment standing in the hood may alter this movement and increase air turbulence. If the emission source is placed near the hood face, vapour is likely to spill outside the hood. Although a high capture or face velocity may help prevent this spillage, a more satisfactory and less energy-consuming approach is to place sources of emission farther into the hood (Figure 8). In some laboratories, a coloured stripe is painted on the hood work surface 15 cm back from the face to serve as a reminder. The less apparatus and fewer bottles the hood contains, the more likely it is to have laminar air flow across its working surface. Observing these simple precautions will often result in a significant improvement in performance. Because a substantial amount of energy is required to supply heated or cooled input air even to a small hood, the use of hoods for the storage of bottles of toxic or corrosive chemicals is a very wasteful practice. This practice may also, as noted above, seriously impair the effectiveness of the hood as a local ventilation device. Thus, it is preferable to provide separate vented cabinets for the storage of toxic or corrosive chemicals. The amount of air exhausted by a small cabinet is much less than that exhausted by a properly operating hood.

The position and movement of the hood user will also affect its performance. A user standing in front of an open hood sash may cause considerable turbulence and eddy currents near its face. Placing equipment well back in the hood and partially closing the sash will help minimize performance losses caused by air turbulence.

After the face velocity of the hood has been measured and the air flow balanced, it should be fitted with a simple manometer or other pressure measuring device (or a velocity measuring device) to enable the user to determine that the hood is operating satisfactorily, to specification. This pressure measuring device should be capable of measuring pressure differences in the range 0.2-5.0 cm of water and should have the low pressure side connected to the duct just above the hood, with the high-pressure side open to the general laboratory atmosphere. Once such a device has been calibrated by measuring the hood face velocity, it will serve as a constant monitor of performance and provide a warning of inadequate performance which might arise from changes in overall laboratory ventilation.

An indirect method of evaluating hood performance is to measure worker exposure while it is being used for its intended purpose. Worker exposure, either peak or time-weighted average, can be measured with commercial personal air-sampling devices worn by the user. However, for

Fig. 10 Positive-pressure glove box for work with oxygen- or moisture-sensitive substances.

most purposes, the combination of a smoke test and a series of face velocity measurements should be sufficient to evaluate hood performance.

3.3. OTHER LOCAL EXHAUST SYSTEMS

The common type of laboratory hood depends on the horizontal flow of a substantial volume of air to direct contaminated air away from the user. Canopy hoods and similar devices use vertical airflow and are much less effective, but they can be useful in providing local ventilation above chromatographic and distillation equipment and various instruments (e.g., atomic absorbtion spectrometers) that cannot reasonably be placed in hoods. With good design, these vertical airflow devices can be used to contain emissions of hazardous substances. Drop curtains or partial walls of plastic or metal to direct airflow may make them more effective. However, without proper training, the users of these devices may find themselves in a contaminated air stream.

3.4. SPECIAL VENTILATION AREAS

3.4.1. Glove boxes and isolation rooms

Glove boxes are usually small units that have multiple ports in which arm-length rubber gloves are mounted, through which the operator works (Figure 10). Construction materials vary widely, depending on the intended use. Clear plastic is frequently used because it allows visibility of the work area and is easily cleaned.

Glove boxes generally operate under negative pressure, so that any air leakage is into the box. If the material being used is sufficiently toxic to require the use of an isolation system, it is obvious that the exhaust air will require special treatment before release into the regular exhaust system. However, because these small units have a low airflow, scrubbing or adsorption (or both) can be accomplished with little difficulty.

Some glove boxes operate under positive pressure. These are commonly used for experiments in which protection of sensitive materials from atmospheric moisture or oxygen is desired. If such glove boxes must be used with materials that present a high toxicity hazard, they should be thoroughly tested for leaks before each use and there should be a method for monitoring the integrity of the system, such as a shut-off valve or a pressure gauge.

Isolation rooms use the same principles as glove boxes, except that the protected worker is within the unit. The unit itself operates under negative pressure, and the exhaust air requires special treatment before release. Many isolation units have a separate air supply to prevent cross-contamination. The workers enter the unit through clean rooms in which they remove their outer clothes, which are replaced by clean work clothes and other personal protective equipment, such as supplied air respirators

Laboratory ventilation 61

(Section 2.4.2). They reverse this procedure when leaving the isolation unit by removing their work clothes in a dirty room, passing through a shower, and then entering a clean room where they collect their outer clothes.

Isolation units should be used only for highly toxic substances that have physical properties that make control difficult or impossible by conventional methods. Other substances should be handled by general safety procedures.

3.4.2. Environmental rooms

Environmental rooms, such as refrigeration cold rooms or warm rooms for culture of organisms and cells, are closed air circulation systems. Thus, the release of any toxic substance in these areas poses potential dangers. Also, because of the contained atmosphere, there is potential for the creation of aerosols and cross-contamination of research materials. These problems should be controlled by preventing the release of aerosols or gases into the room environment.

Because of the contained atmosphere in environmental rooms, provision should be made to allow workers to escape rapidly, if necessary. The doors should be equipped with magnetic latches or breakaway handles that allow a trapped person to open the door. The electrical system should be independent from the main power supply so that such workers are not confined in the dark.

Volatile flammable solvents must not be used in cold rooms because the exposed motors of the circulating fans can serve as a source of ignition and initiate an explosion. The use of volatile acids must also be avoided because these can corrode the cooling coils in the refrigeration system, which can lead to leaks of refrigerant.

3.5. MAINTENANCE OF VENTILATION SYSTEMS

Even the best engineered and installed ventilation system requires routine maintenance. Blocked or plugged air intakes and exhausts may alter the performance of the total ventilation system. Belts loosen, bearings require lubrication, motors need attention, ducts corrode, and minor components fail. These malfunctions, individually or collectively, can affect overall ventilation performance.

All ventilation systems should have a monitoring device that readily permits the user to determine whether the total system and its essential components are functioning properly. Manometers, pressure gauges, and other devices that measure the static pressure in the air ducts are sometimes used to reduce the need for manual measuring of the airflow. The need for monitoring devices and their type should be determined on a case by case basis. If the substance being contained has an intrinsic warning property such as a strong odour and the consequence of overexposure is minimal, the system will need less stringent control than if the substance is highly toxic or has poor warning properties. The need for scheduled maintenance will also be determined by these factors.

4

Design Requirements and Use of Electrically-powered Laboratory Apparatus

During the past 25 years, the use of electrically powered apparatus in laboratories has increased more rapidly than that of any other category of equipment. Such equipment is now used routinely for operations requiring heating, cooling, agitation or mixing, and pumping, as well as for a large variety of instruments used in making physical measurements. In fact, electrical apparatus is now so commonly available that it should be the only type of heat source used in areas where there are flammable materials. However, although the introduction of electrically powered equipment has resulted in major improvements in laboratory safety, the use of this equipment does pose a new set of possible hazards.

4.1. GENERAL PRINCIPLES

Laboratory workers should know the procedures for removing a person from contact with a live electrical conductor and the emergency first aid procedures to use for a person who has received a serious electrical shock (Section 2.6.5).

All 110-V or 220-V single-phase outlet receptacles in laboratories should be of a standard design that accepts a three-prong plug and provides a ground or "earth" connection. The use of an old style two-prong receptacle and an adapter that takes a three-prong plug is an unsatisfactory alternative. No attempt should be made to bypass the connection to ground. Old-style receptacles should be replaced as soon as possible, and an additional ground wire added if necessary. The use of an extension cord should be avoided, but if this is not possible, a standard three-conductor extension cord that provides an independent ground connection should be used.

Receptacles that provide electric power for operations in hoods should be located outside the hood. This prevents the formation of electrical sparks inside the hood when a device is plugged in, provided that the cord is first connected to the device in the hood and then to the power receptacle. Such receptacles also allow a laboratory worker to disconnect electrical devices from outside the hood. However, they should not be positioned in such a way that they may be pulled, get in the way of equipment or come into contact with corrosive chemicals.

Laboratory equipment that is to be plugged into a 110-V or 220-V receptacle should be fitted with a standard three-conductor line cord that provides a proper ground connection to the chassis of the apparatus; the cord serving as the electrical conductor should have a male plug which fits the receptacle without need for an adaptor. All frayed or damaged line cords should be replaced before further use of the equipment is permitted. Annual inspection of all electrical cords is good practice. It is also desirable that equipment that will be plugged into an electrical receptacle be fitted with a fuse of appropriate rating or other overload protection device that will disconnect the electrical circuit if the apparatus fails or is overloaded. This overload protection is particularly useful for apparatus likely to be left on and unattended for long periods, such as variable auto-transformers, vacuum pumps, ovens, motors, and electronic instruments.

Motor driven electrical equipment used in a laboratory where volatile flammable materials may be present should be equipped with a non-sparking induction motor rather than with a series-wound motor that uses carbon brushes. This precaution applies to the motors used in vacuum pumps, mechanical shakers, stirrer motors, magnetic stirrers, and rotary evaporators. The speed of an induction motor operating under load should not be controlled by a variable auto-transformer, which might cause the motor to overheat and start a fire. It is impossible to modify a series-wound motor to make it spark-free. For this reason, kitchen appliances, such as mixers or blenders, should not be used in laboratories where flammable materials may be present. Finally, it must be remembered that when equipment with series-wound motors, such as a vacuum cleaner or portable electric drill, is brought into a laboratory the absence of flammable vapours should be ensured before the equipment is used.

Electrical equipment should be located so as to avoid the possibility of water or chemicals being spilled on it. If water or a chemical should accidentally be spilled on electrical equipment, the equipment should be unplugged immediately at the power source and not used until it has been cleaned and inspected, preferably by a qualified technician. Water can also enter electrical equipment placed in a cold room or refrigerator through condensation. Cool rooms pose a particular hazard (Section 3.4.2) because their atmosphere may be at a high relative humidity. If electrical equipment must be used in such areas, condensation must be lessened, but not eliminated, by mounting the equipment on a wall or vertical panel. The potential for electrical shock in these rooms can be minimized by careful electrical grounding of the equipment.

Repairs and modifications to electrical equipment should be carried out by a qualified technician, remembering that except for certain instrument adjustments, line cords of electrical equipment should always be unplugged at the power source before adjustments, modifications, or repairs are undertaken. When it is necessary to use a piece of electrical equipment that is plugged in, laboratory workers should be certain that their hands are dry and that they are not standing on a damp floor.

4.2. VACUUM PUMPS

Vacuum distillations that involve relatively volatile substances should normally be performed by using a water or steam aspirator, rather than a mechanical vacuum pump. However, the distillation of less volatile substances, removal of final traces of solvents, and some other operations require pressures lower than can be obtained with a water aspirator and are normally performed by mechanical vacuum pump (Section 8.3). The input line from the system to the vacuum pump should be fitted with a cold trap to collect volatile substances from the system and minimize the quantity that dissolves in the pump oil. The use of liquid air or nitrogen in such traps, however, can lead to a flammability hazard (Section 8.3.4).

The exhaust from a mechanical pump should be vented to a suitable air exhaust system, for example by connecting a flexible tubing from the pump exhaust port to the back of a hood. It may also be necessary to scrub or absorb vapours from the pump exhaust if they are toxic. Even with good practices, volatile toxic or corrosive substances may accumulate in the pump oil and be discharged into the laboratory atmosphere during subsequent pump use. This hazard should be avoided by draining and replacing the pump oil when it becomes contaminated. The contaminated pump oil should be disposed of by following an acceptable standard procedure for the safe disposal of toxic or corrosive substances (Part III).

Belt-driven mechanical pumps with exposed belts should have protective guards. Such guards are particularly important for pumps installed on portable carts or the tops of benches where laboratory workers or others might accidentally entangle clothing or fingers in the moving belt.

4.3. DRYING OVENS

Electrically-heated ovens are commonly used in the laboratory to remove water or other solvents from chemical samples and to dry laboratory glass- or plastic-ware. With the exception of vacuum drying ovens, ovens rarely have any provision for preventing the discharge of the volatile substances into the laboratory atmosphere. Thus, it must be assumed that these substances will escape into the laboratory atmosphere; organic substances could also be present in concentrations sufficient to form explosive mixtures with the air inside the oven.

Ovens should not be used to dry any chemical sample that has even moderate volatility and might pose a hazard because of acute or chronic toxicity, unless special precautions have been taken to ensure continuous venting of the atmosphere inside the oven. Thus most organic compounds should not be dried in a conventional laboratory oven. This should include silica gel which is known to have adsorbed organic material.

Glassware that has been rinsed with an organic solvent should not be dried in an oven. If rinsing is necessary, the item should be rinsed successively with a water-soluble organic solvent and purified water before being placed in the oven.

Because of the possible formation of explosive mixtures between volatile substances and the air inside an oven, laboratory ovens should be constructed so that their heating elements (which may become red hot) and their temperature controls (which may produce sparks) are physically separated from their interior atmospheres. Many small household ovens and similar heating devices do not meet these requirements and should not be used in laboratories. Existing ovens that do not meet these requirements should either be modified or have a prominent sign attached to the oven door to warn workers that flammable materials should not be placed in that oven. Some safety groups suggest that every laboratory oven should be modified by placing a blowout panel in its rear wall so that explosion within the oven will not blow the door and the contents into the laboratory. The area behind the panel should be able to withstand such a blowout.

Mercury thermometers should not be mounted through holes in the tops of ovens so that the bulb is unprotected. Bimetallic strip or electronic thermometers are the preferred alternative. If a mercury thermometer is broken in an oven, the oven should be turned off, cooled, and stripped down so that all the mercury can be removed with due precautions.

4.4. REFRIGERATORS

The potential hazards posed by laboratory refrigerators are similar to those of drying ovens. As there is no simple and satisfactory method for continuously venting the interior atmosphere of a refrigerator, any vapours escaping from vessels in the refrigerator will accumulate. Thus, the atmosphere in a refrigerator could contain an explosive mixture of air and flammable vapour or a dangerously high concentration of toxic vapour, or both. The problem of toxicity could be aggravated by laboratory workers who place their faces inside the refrigerator while searching for a particular sample, thus inhaling some of the atmosphere. Laboratory refrigerators should never be used for the storage of food or beverages (Section 1.3), nor should laboratory ice be used in preparing drinks.

There should be no potential source of electrical sparks on the inside of a laboratory refrigerator. When purchasing a refrigerator for a laboratory it is preferable to select one that has been designed for "flammable storage". If this is not possible, new or existing refrigerators should be modified by removing all spark sources from the refrigerated compartment, i.e., (a) remove any interior light activated by a switch mounted on the door frame, (b) move the contacts of the temperature control thermostat to a position outside the refrigerated compartment, and (c) move the contacts for any fan control thermostat from the refrigerated compartment to the outside. "Frost-free" refrigerators are not advisable for laboratory use because of the problems of modifying them for safe laboratory use. Many have a drain tube or hole that carries water and flammable liquid to an area adjacent to the compressor, thus presenting a spark hazard. The electric heaters used to defrost the freezing coils are also a potential hazard.

Laboratory refrigerators should be placed against fire-resistant walls: they should have heavy duty electrical cords, and preferably be protected by their own circuit breakers.

Uncapped containers of chemicals should never be placed in a refrigerator. Chemicals should be capped in such a way as to achieve a seal that is both vapour tight and unlikely to permit a spill if the container is tipped over. Aluminium foil caps, corks, corks wrapped with aluminium foil, or glass stoppers do not usually meet these criteria. Such methods for capping containers should be discouraged. The most satisfactory temporary seals are achieved by using containers with screw caps lined with either a conical polyethylene or a Teflon® insert. The best containers for long-term storage are sealed, nitrogen-filled glass ampoules.

Placing potentially explosive or highly toxic substances in a laboratory refrigerator must be discouraged. If such substances have to be stored, a clear, prominent warning sign should be placed on the outside of the refrigerator door. Each sample placed in a refrigerator should be labelled with its contents, owner, and date of acquisition, preparation, or storage. The period of storage of such material in the refrigerator should be kept to a minimum.

4.5. STIRRING AND MIXING DEVICES

The stirring and mixing devices commonly found in laboratories include electrically driven stirrer motors, magnetic stirrers, shakers, small pumps for fluids, and rotary evaporators for solvent removal. Water-driven stirrers and pumps are sometimes also used. These devices may be used in laboratory operations that are performed in a hood, and it is important that they be operated to preclude the production of electrical sparks within the hood. Furthermore, it is important that, in the event of an emergency, a laboratory worker can turn such devices off from outside the hood. Finally, heating baths used in association with these devices should also be spark-free and capable of control from outside the hood.

Only spark-free induction motors should be used to run stirring and mixing devices in a hood. Any motor that produces sparks during start-up or operation should not be used. Fortunately, many modern motors are satisfactory in this respect, but the on/off switches and rheostat type speed controls of some motors are not. The switch or rheostat has exposed contacts that can produce an electrical spark when a change in the setting is made. This is particularly so for magnetic stirrers and rotary evaporators. One solution may be that a competent electrician inserts a switch in the line cord near the plug end, that is outside the hood, to avoid the dangers arising. The speed of an induction motor operating under load should not be controlled by a variable auto-transformer (Section 4.1).

Stirring and mixing devices are often operated for long periods without continual attention (e.g., reaction mixtures that are stirred overnight). Therefore the consequences of stirrer failure, electrical overload, or impairment of the motion of the stirring impeller should be considered. It is good practice to attach a stirring impeller to the shaft of the stirrer motor by lightweight rubber tubing. Then, if the motion of the impeller becomes blocked (e.g., by formation of a copious precipitate), the rubber simply twists until it breaks rather than the motor stalling or the glass apparatus breaking. It is also desirable that an unattended stirring motor be fitted with a suitable fuse or overload-protection device.

4.6. HEATING DEVICES

The most common electrical devices in a laboratory are those that supply the heat for reactions and separations. These include hotplates, heating mantles and tapes, oil baths, air baths, hot-tube furnaces, and hot-air guns. Although much safer than gas burners, they can pose electrical and fire hazards if used improperly.

> 1. The heating element in a laboratory heating device should be enclosed in a glass, ceramic, or insulated metal case so that it is not possible for a worker or a metallic conductor to touch the wire carrying the electric current. This practice minimizes the likelihood of electrical shock and of accidentally producing an electric spark near a flammable liquid or vapour. This type of construction diminishes the possibility of a flammable liquid or vapour coming in contact with the hot wire, whose

temperature is higher than the ignition temperature of many common solvents. If any heating device becomes so worn or damaged that its element is exposed, the device should either be discarded or repaired. Note that many household appliances (e.g., hot-plates and space heaters) do not have enclosed heating elements and are not suitable for laboratory use.

2. The temperature of many laboratory heating devices (e.g., heating mantles, air baths, and oil baths) is controlled by a variable auto-transformer that supplies a fraction of the total line voltage to the heating element. Variable auto-transformers should be wired with grounded three-prong plugs in such a manner that both wires of the output from the power line are disconnected when the switch is in the off position. If this method is not used, each output line may be at a relatively high voltage with respect to ground and capable of delivering a lethal electric shock. The outer cases of all variable auto-transformers have numerous openings to allow for ventilation; since sparking may occur when the voltage is adjusted, laboratory workers should be careful to locate these devices where water and other chemicals cannot be spilled on them (shock hazard), and where their movable contacts will not be exposed to flammable liquids or vapours (fire hazard). Specifically, variable auto-transformers should be mounted on walls or vertical panels outside hoods. They should not be placed on laboratory bench tops, especially those inside hoods.

3. Electrical input lines (even lines from variable transformers) to almost all laboratory heating devices may be at a potential of 110-V or 220-V with respect to electrical ground, and should therefore be considered potential shock and spark hazards. Thus, any connections to heating devices should be both mechanically and electrically secure and completely covered with an insulating material. Alligator (crocodile) clips should not be used to connect a line cord from a variable auto-transformer to a heating device (especially an oil bath or an air bath). All connections should be made by using either insulated binding posts or, preferably, a plug and receptacle combination.

4. Whenever an electrical heating device is to be left unattended for a significant period of time (e.g., overnight), it is advisable to equip it with a temperature sensor that will turn off the electric power if the temperature of the heater exceeds a preset limit. Similar controls are available for turning off the electric power if the flow of cooling water through a condenser is ter-

minated or reduced. Such fail-safe devices prevent fires or explosions that may occur if the temperature of an unattended reaction increases significantly because of a change in line voltage or because reaction solvent is lost. Steam (rather than electrically-heated) devices are generally preferred whenever temperatures of 100 °C or less are required. These devices do not present shock or spark hazards and can be left unattended as the temperature will not exceed 100 °C.

4.6.1. Laboratory hot-plates

Laboratory hot-plates are often used when solutions are to be heated above 100 °C. As noted above, only hot-plates that have completely enclosed heating elements should be used in laboratories. Although most modern hot-plates meet this criterion, many older ones pose an electrical spark hazard from the on-off switch or the bimetallic thermostat used to regulate the temperature. A warning label should be put on any spark-prone hot-plate, and no hot-plates should be purchased that do not have spark-free construction. It is often not obvious when a hot-plate or heating block is on. To prevent skin burns, a warning should be posted by such devices when they are switched on (Figure 11).

4.6.2. Heating mantles

Heating mantles are commonly used for heating round-bottomed flasks, reaction kettles, and other vessels. These are constructed by enclosing the heating element in a series of layers of fibreglass cloth. As long as the fibreglass coating is neither worn nor broken, and no water or other chemical is spilled onto the mantle, they pose no shock hazard.

Heating mantles are always intended to be used with a variable auto-transformer to control the input voltage and should never be plugged directly into a 110-V or 220-V line. Laboratory workers should be careful not to exceed the input voltage recommended by the manufacturer because higher voltages will cause the mantle to overheat, melting the fibreglass insulation and exposing the heating element. In general, the maximum recommended input voltage for a mantle that is being used with a dry flask is 10-20 V lower than that for a mantle that is being used with a flask containing a liquid. If a metal-enclosed mantle is used, it is good practice to ground the outer case either by using a three-conductor cord (containing a ground wire) from the variable auto-transformer or by attaching one end of a heavy gauge conductor to the mantle case and the other end to a good electrical ground such as a main cold water pipe. This practice provides the laboratory worker with protection against an electric shock if the heating element inside the mantle should be shorted against

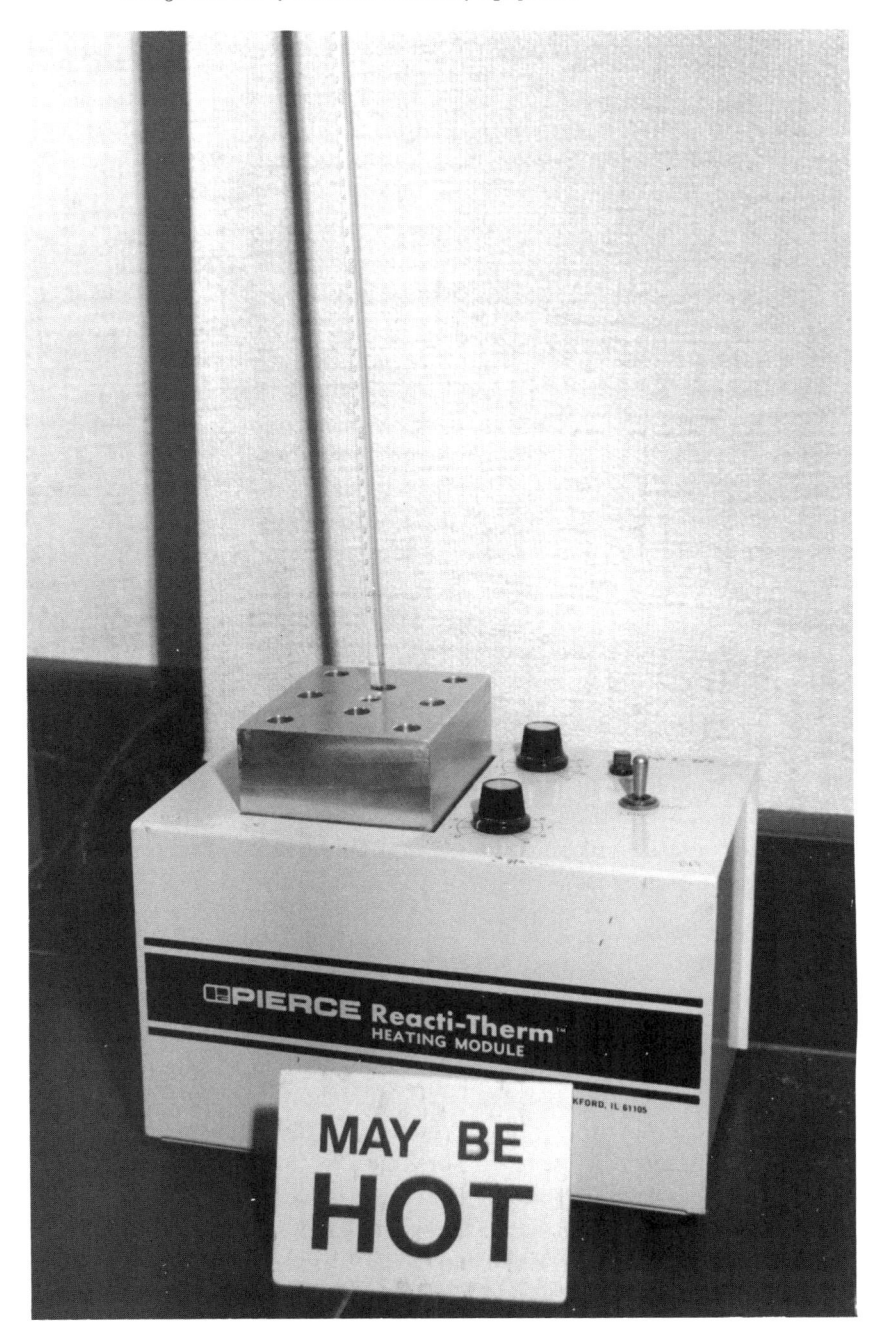

Design and use of electrical laboratory equipment

Fig. 11 Heating block with prominent warning sign to guard against accidental contact with skin or clothing.

the metal case. All glassware in heating mantles should be properly supported (Figure 12).

4.6.3. Oil baths

Electrically-heated oil baths are often used as heating devices for small or irregularly shaped vessels or when a stable heat source is desired that can be maintained at a constant temperature. For temperatures below 200 °C, a saturated paraffin oil is often used. A silicone oil (which is more expensive) should be used for temperatures up to 300 °C. Such baths should be used in hoods for protection against any fumes that may be emitted. An oil bath should always be monitored by a thermometer or other temperature-sensing device to ensure that its temperature does not exceed the flash point of the oil. For the same reason, an oil bath left unattended should be fitted with a temperature-sensing device that will turn off the electric power if the bath overheats. Bare wires should not be used as resistance heaters in oil baths. These baths should be heated with a metal pan fitted with an enclosed heating element or with a heating element enclosed in an insulated metal sheath.

Heated oil should be contained in either a metal pan or a heavy walled porcelain dish. A glass dish or beaker can easily break and spill hot oil. The bath should be carefully mounted on a stable horizontal support such as a laboratory jack that can be easily raised or lowered without danger of the bath tipping over. It should never be supported on an iron ring because of the greater likelihood of spillage. Finally, a laboratory worker using an oil bath heated above 100 °C should guard against the possibility of water or some other volatile substance falling into the hot bath, which can spatter hot oil over a wide area.

4.6.4. Air baths

Electrically-heated air baths may be used as a substitute for heating mantles when heating small or irregularly shaped vessels. Because of their inherently low heat capacity, such baths normally must be heated to 100 °C or more above the desired temperature of the vessel. They should be constructed so that the heating element is completely enclosed and the connection to the air bath from the variable transformer is mechanically and electrically secure. They may be constructed from metal, ceramic, or (less desirably) from glass. If a glass vessel is used, it should be thoroughly wrapped with a heat-resistant tape so that if the vessel is broken, the glass will be retained and the bare element will not be exposed.

4.6.5. Heat guns

Laboratory heat guns are constructed with a motor-driven fan that blows air over an electrically-heated filament. They are frequently used to dry glassware or to heat the upper parts of a distillation apparatus

Design and use of electrical laboratory equipment 73

Fig. 12 Round-bottomed flask, with fittings, properly supported over a heating mantle.

during distillation of high-boiling materials. The heating element becomes red hot during use and cannot be enclosed, and the on/off switches and fan motors are not usually spark-free. For these reasons, heat guns pose a serious spark hazard. They should never be used near open containers of flammable liquids, nor in environments where appreciable concentrations of flammable vapours may be present, nor in hoods that are being used to remove flammable vapours. Household hair dryers may be used as substitutes for laboratory heat guns only if they have three-conductor or double-insulated line cords.

4.7. ELECTRONIC INSTRUMENTS

Most modern electronic instruments are fitted with a line cord that contains a separate ground wire for the chassis and with a suitable fuse or other overload protection. Any existing instrument that lacks these features should be modified to incorporate them. As with any other electrical equipment, precautions should be taken to avoid the possibility of water or other chemicals being spilled on to the instrument.

Laboratory workers should not undertake repairs, alterations, or adjustments to electronic instruments without supervision unless they have received special prior instruction. This precaution is particularly important with instruments that incorporate high voltage circuits, such as oscilloscopes, electrophoresis equipment, spectrometers that have photomultiplier tubes, or equipment that uses vacuum or discharge tubes.

Part II

Safe Storage and Use of Hazardous Chemicals

5

Procedures for the Procurement, Storage and Distribution of Chemicals

5.1. ORDERING AND PROCUREMENT OF CHEMICALS

The safe handling, use, and disposal of hazardous substances begins with the personnel who requisition such substances and those who approve the purchase orders. These people must be aware of the potential hazards of the substances being ordered, know whether or not adequate facilities and trained personnel are available to handle them and should ensure that safe disposal procedures are available.

Before a new hazardous substance is received, information concerning its proper handling, including proper disposal, should be given to those who will be involved in its use. Relevant information is often included in chemical catalogues from laboratory suppliers. If the receipt of a hazardous chemical involves receiving room or storeroom personnel, they should be advised that the substance has been ordered. It is the responsibility of the laboratory supervisor to ensure that the facilities are adequate and that those who will handle the material have received proper training and education to do so safely.

Material safety data sheets, which detail the physical properties and toxicological information for a large number of substances, can be obtained from vendors. However, the quality and depth of information on these sheets may vary widely. For substances for which such sheets are not available, e.g., research chemicals sold in small quantities, the manufacturer will usually provide whatever health and safety information is available.

Shipments of hazardous substances should have labels indicating the hazards of each substance, providing some information for delivery, storeroom, and stockroom personnel. However, this information may not be available once the container has been opened.

As storage areas in laboratories are usually restricted in size, it is sometimes preferable to order small amounts of chemicals to avoid the hazards associated with repackaging by storeroom personnel. There is also less likelihood of over-ordering. Unused chemicals can constitute a significant portion of the hazardous waste generated from laboratories. Much of this waste consists of containers from which only a small fraction of the chemical has been removed; the partly used container may stay on a shelf for years before being disposed of. The cost of storing and disposing of chemicals has risen to the point that it is often more than the cost of the chemical, and much more than any saving that may be made by ordering in bulk.

It is preferable that all substances be received at a central location for distribution to the storerooms, stockrooms, and laboratories. A central receiving area is also helpful in monitoring substances that may eventually enter the waste-disposal system. An inventory of substances kept in the storerooms and stockrooms gives an indication of the quantity and nature of substances that have to be disposed of in the future.

No container of a chemical or cylinder of a compressed gas should be accepted without an identifying label, which should include the following components:

1. Identification of contents.
2. Principal hazard (e.g., flammable liquid) with a brief description.
3. Precautions to minimize hazard and prevent an accident.
4. First aid in case of exposure.
5. Procedures for cleaning up spillage.
6. If appropriate, special instructions to a physician who might be called to an accident.

Delivery, storeroom and stockroom personnel should be trained in the handling of hazardous substances. Training should teach staff how to handle containers of chemicals so that they are not dropped, bumped, or crushed by being piled one upon another. Information must be provided about environmental and exposure hazards to be avoided. This should include the following points:

1. The use of proper material handling equipment, protective apparel (Sections 2.1-2.3), and safety equipment (Section 2.4).
2. Emergency procedures, including the clean-up of spillages (Section 2.5.3), and the disposal of broken containers.

3. The dangers of contact with chemicals by skin absorption, inhalation, or ingestion (Section 6.1).
4. The meaning of the various labels on shipping packages.
5. The proper methods of material handling and storage, especially the incompatibility of some common substances, the dangers associated with "alphabetical" storage, and the sensitivity of some substances to heat, moisture, light, and other storage hazards (See Appendices A-E)
6. The special requirements for handling heat-sensitive materials, including those shipped by refrigeration or packed in solid carbon dioxide.
7. The general and specific problems associated with compressed gases (Sections 5.2.6 and 8.1).
8. The hazards associated with flammable liquids (especially the danger of their vapours catching fire some distance from the container), explosives (Chapter 7), toxic gases and vapours, and oxygen displacement by an inert gas such as nitrogen or carbon dioxide.
9. Substances that react with water to produce dangerous situations (e.g., alkali metals, burning magnesium, metal hydrides, acid chlorides, phosphides, carbides are listed in Appendix A).
10. Regulations governing controlled substances (such as radioactive materials,drugs,ethanol, explosives), and needles and syringes.
11. Chemicals that have offensive smells.
12. Packages that exhibit evidence that the inside container has broken and leaked its contents.

5.2. STORING CHEMICALS IN STOREROOMS AND STOCKROOMS

The requirements for the storage of chemicals vary according to the size of the organization, the quantities to be handled, and the nature of their hazards.

Often, the provision of adequate storage space is given little consideration in the design of laboratory buildings. Lack of sufficient storage space can create hazards due to overcrowding, storage of incompatible chemicals together, and poor housekeeping. Properly designed and ventilated facilities of adequate size should be provided to ensure the safety of personnel and the protection of the buildings.

In many institutions, chemicals are sent directly to the individual who initiated the order. If the facilities of that individual's laboratory are appropriate for the materials being used, this system may be satisfactory. However, experience has shown that it is usually necessary to maintain a

80   *Chemical Safety Matters*

reserve of supplies in excess of the amounts that can be kept safely in the laboratory. If the quantities are large or the volumes of the individual containers are such that repackaging is necessary, then a storage area is required. This could be a stockroom in a laboratory building or a central storeroom serving the whole institution.

Stored chemicals should be examined periodically, at least annually. Those that have been kept beyond their shelf life or have deteriorated, have questionable labels, are leaking, have corroded caps, or are potentially unsafe for any other reason should be disposed of in a safe manner. A first-in, first-out, system of stock keeping should be used.

Chemicals on shelves should be prevented from falling off by placing retaining cords or similar restraining devices across the open face of the shelf or by raising the forward face of the shelf (Figure 13).

5.2.1.   Stockroom design

Stockrooms are similar to central storerooms except that the quantities of materials held are usually much smaller, and such rooms are usually within or close to the areas served.

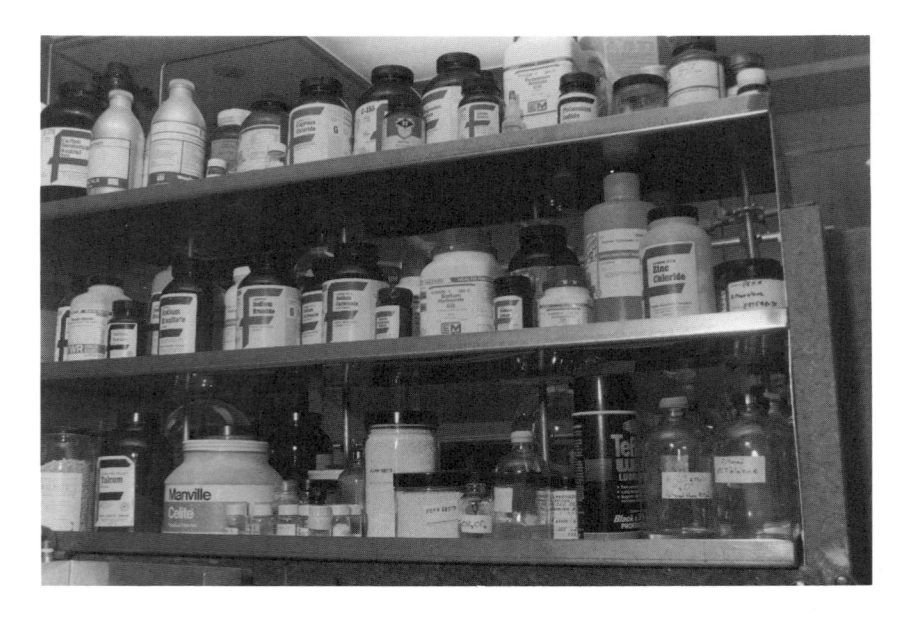

Fig. 13   Small bottles of reagents and other materials for day-to-day use stored on secure shelving.

Stockrooms should not be used as preparation areas, preparation and repackaging should be performed elsewhere.

Stockrooms should be conveniently located and open during normal working hours so that laboratory workers need not store excessive quantities of chemicals. However, this does not imply that all laboratory workers should have unlimited access to the chemicals in the stockroom. Procedures must be established for the operation of any stockroom that places responsibility for its safety and inventory control in the hands of one person. If it is not feasible to have a full-time stockroom clerk, then one person who is readily available should be assigned that responsibility.

Stockrooms should be well ventilated. If storage of opened containers is permitted, extra local exhaust ventilation and the use of outside storage containers or spill trays are necessary.

5.2.2. Flammable liquids

Centralized storage of bulk quantities of flammable liquids provides the best method of controlling fire hazards. The most effective way to minimize the impact of a hazard is to isolate it, and therefore storage and dispensing rooms for flammable liquids are best located in a special building separated from the main building. If this is not feasible, and the area must be located in a main building, the preferred location is a "cut-off" area at ground level, having at least one exterior wall. Under no circumstances should storage rooms for flammable liquids be located on the roof, or an upper floor, or below ground level, or in the centre of the building. All of these locations are undesirable because they are less accessible for fire-fighting and potentially dangerous to people in the building.

"Cut-off" is a fire protection term indicating that the walls, ceilings, and floors of an inside storage room for flammable liquids should be constructed of materials with at least a 2-hour fire resistance, and the room should be fitted with self-closing fire doors. All storage rooms should have adequate mechanical ventilation controlled by a switch outside the door and be fitted with explosion-proof lighting and switches. Other potential sources of ignition, such as burning tobacco and lighted matches, should be forbidden.

5.2.3. Drum storage

Two-hundred-litre drums are commonly used to ship flammable liquids, but are not intended as long-term storage containers. It is unsafe to dispense from sealed drums in the state that they are received. The bung should be removed carefully and replaced by an approved pressure relief vent to protect against internal build-up pressure from heat from fire or exposure of the drum to direct sunlight.

Drums should be stored on metal racks placed so that the end bung faces an aisle and the side bung is on top. The drums, as well as the racks, should be grounded with an appropriate length of heavy gauge copper wire and spring type battery clamps. Because effective grounding requires metal-to-metal contact, all dirt, paint, and corrosion must be removed from the contact areas. It is also necessary to connect metal receiving containers to the dispensing drum with similar clamps and wire. Drip pans that have flame arresters should be installed or placed under taps or faucets.

Dispensing from drums is usually done by one of two methods.

The first, and safer method, is to dispense liquids from a vertical drum with a hand operated rotary transfer pump. Such pumps have metering options and permit immediate cut-off control to prevent overflow and spillage. These can be reversed to siphon off excess liquid in case of overfilling, so as to return it to the drum.

The second method involves gravity feed through drum faucets that are self-closing and require constant hand pressure for operation. Plastic faucets are not generally acceptable because of chemical action on them.

5.2.4. Toxic substances

Toxic substances should be segregated from other substances and stored in an area that is cool, well ventilated, and away from light, heat, acids, oxidizing agents, moisture, etc.

The storage of unopened containers of toxic substances normally presents no major problem. However, because containers may develop leaks or suffer breakages, storerooms should be equipped with exhaust hoods (Section 3.2) or equivalent local ventilation devices in which toxic substances can be handled.

Opened containers of toxic substances should be closed with tape or other sealant before being returned to the storeroom. Section 6.9 describes storage and record keeping procedures for compounds of known high chronic toxicity.

Volatile carcinogens, in suitable containers, should be stored in spark-free freezers (-20 $^{\circ}$C) to avoid evaporation. Such compounds should be labelled, and placed in a hood to allow them to reach room temperature before use.

5.2.5. Water-reactive chemicals

Some chemicals react with water to evolve heat and flammable or explosive gases. For example, potassium and sodium metals and many metal hydrides react with water to produce hydrogen, and these reactions often evolve sufficient heat to ignite the hydrogen with explosive violence. Certain polymerization catalysts, such as aluminium alkyls, react and burn violently on contact with water. Other water-reactive chemicals are listed in Appendix A.

Storage facilities for water-reactive chemicals should be constructed to prevent their accidental contact. This is best accomplished by eliminating all sources of water in the storage area. Automatic sprinkler systems should not be installed where large quantities of water-reactive chemicals are stored. Facilities should be of fire-resistant construction, and other combustible materials should not be stored in the same area.

5.2.6. Compressed gases

Cylinders of compressed gases should be stored in well ventilated, dry areas. Where practicable, these should be fire-resistant and at ground level. Cylinders may be stored outdoors, but protection must be provided to prevent corrosion, and air circulation must not be restricted. Local regulations may specify the storage condition to be used.

Compressed gas cylinders should not be stored near sources of ignition or where they might be exposed to corrosive chemicals or vapours. Nor should they be stored where heavy objects might strike or fall against them, such as near elevators (lifts or hoists), service corridors, and unprotected platform edges.

The cylinder storage area should be posted with the names of the gases stored. Where gases of different types are stored at the same location, the cylinders should be grouped by type of gas (e.g., flammable, toxic, or corrosive). Flammable gases should be stored separately from other gases, and provision should be made to protect them from fire. Full and empty cylinders should be stored in separate parts of the storage area and arranged so that older stock can be used first, with minimum handling of other cylinders.

Cylinders and valves are usually equipped with safety devices, including a fusible metal plug that melts at a relatively low temperature (70-95 $^{\circ}$C). Although most cylinders are designed for safe use up to a temperature of 50 $^{\circ}$C, they should not be placed near radiators, steam pipes, or boilers not in direct sunlight. Cylinder caps should be in place at all times during storage and movement to and from the store.

Stored cylinders should be secured in an upright or horizontal position. Acetylene cylinders should always be stored valve end up to minimize the possibility of solvent discharge (Section 8.4.1). Oxygen should be stored in an area that is at least 6 m away from any flammable or combustible materials (including oil and grease) or separated from them by a non-combustible barrier at least 1.5 m high and having a fire resistance rating of at least 30 minutes.

Cylinders are sometimes painted by the vendor to aid recognition of their contents, but this colour coding is not always reliable and can vary nationally. The name stencilled or printed on the cylinder is the only acceptable method of identification. If it is suspected that a stored cylinder is leaking, the procedures described in Section 10.5 should be followed.

5.3. DISTRIBUTING CHEMICALS FROM STOCKROOMS TO LABORATORIES

The method of transporting a chemical between stockrooms and laboratories must reflect the potential danger posed by the specific substance.

5.3.1. Chemicals

When chemicals are carried by hand, they should be placed in a box or acid-carrying bucket to protect against breakage and spillage. When they are transported on a wheeled cart, the cart should be stable under the load and have wheels large enough to negotiate uneven surfaces (such as expansion joints and floor drain depressions) without tipping or stopping suddenly.

If possible, chemicals should be carried upstairs or transported on freight elevators to avoid exposure to other persons on passenger elevators. However, it is unwise to carry solid carbon dioxide in any closed elevator.

Provisions for the safe transport of small quantities of flammable liquids include: the use of rugged, pressure-resistant, non-venting containers; storage during transport in a well ventilated vehicle; and elimination of potential ignition sources.

5.3.2. Cylinders of compressed gases

Cylinders that contain compressed gases should not be subjected to rough handling or abuse. Such misuse can seriously weaken the cylinder, rendering it unfit for further use or causing a highly dangerous fracture. To protect the valve during transportation, the cover cap should be screwed on tightly by hand and remain on until the cylinder is in place and ready for use. Cylinders should never be rolled or dragged. The preferred transport, even for short distances, is by hand truck or cart with the cylinder strapped in place. Only one cylinder should be handled at a time.

5.4. STORING CHEMICALS IN LABORATORIES

The quantities of toxic, flammable, unstable, or highly reactive materials that are permitted in laboratories should be controlled. Restricting quantities arbitrarily may interfere with laboratory operations; unrestricted quantities can result in undesirable accumulation of such materials in the laboratory. Hence it is necessary to balance the needs of the laboratory workers and the established requirements for safety. Such decisions will be affected by the level of competence of the workers, the

Fig. 14 Storage of bottles of corrosive liquids in secondary containers or on catch trays.

level of safety features designed into the facility, the location of the laboratory, the nature of the chemical operations, and the accessibility of the stockroom. In some cases, local regulations or insurance requirements will determine the quantities that can be stored. In general, all laboratories should have two exits (one may be an emergency exit) so that a fire at one exit will not block occupants' escape. Doors that open outward are desirable.

5.4.1. General considerations

Every chemical in the laboratory should have a definite storage place and should be returned to that location after use.

The storage of chemicals on bench tops is undesirable. In such locations, they are unprotected from potential exposure to fire and are also more readily knocked over. Storage in hoods is also inadvisable because this practice interferes with the air flow in the hood (Section 3.2, 3.3), reduces the working space, and increases the amount of material that could become involved in a hood fire.

Storage trays or secondary containers should be used to minimize the spillage of material should a container break or leak (Figure 14). Most laboratory workers tend to store hazardous materials in the cabinet space under hoods, so that ventilation of these cabinets is advisable. The use of such cabinets also has the advantage that, because of the proximity to the hood, the safe practice of transferring hazardous materials in the hood itself is encouraged.

As in storerooms and stockrooms, care should be taken in the laboratory to avoid exposure of chemicals to heat or direct sunlight, and the proximity of incompatible substances (Appendix E).

Laboratory refrigerators are to be used solely for the storage of chemicals. Food must not be placed in them. All containers placed in the refrigerator should be properly labelled with contents and owner, date of acquisition or preparation, and nature of any potential hazard. If necessary, they should be sealed to prevent the escape of any corrosive vapours. Flammable liquids must not be stored in refrigerators unless the unit is an approved, explosion-proof, or laboratory-safe type (Section 4.4).

An inventory of chemicals stored in the laboratory should be made periodically. Unneeded items should be returned to the stockroom or storeroom. At the same time, containers that have illegible labels and chemicals that appear to have deteriorated should be disposed of in the proper manner (Part III).

When a worker moves out of a laboratory, he or she and the laboratory supervisor should arrange for the removal or safe storage of all hazardous materials left behind.

5.4.2. Flammable liquids

Quantities of flammable liquids greater than 1 litre are best stored in metal containers. Portable approved safety cans are the safest vessels for storing flammable liquids. These cans are available in a variety of sizes and materials. They have spring-loaded spout covers that can open to relieve internal pressure when subjected to heat and will prevent leakage if tipped over. Some are equipped with a flame arrester in the spout that will prevent flame propagation into the can. If possible, flammable liquids received in large containers should be repackaged into safety cans for distribution to laboratories (Section 5.2.3). Such cans must be properly labelled to identify their contents.

Small quantities of flammable liquids should be stored in ventilated storage cabinets made of heavy gauge steel with riveted and spot-welded seams. Such cabinets are of double wall construction and have a 4 cm air space between the inner and the outer walls. The door is 5 cm above the bottom of the cabinet, and the cabinet is liquid-tight to this level. It is provided with vapour-venting provisions and can be equipped with a sprinkler system. Materials that react with water should not be stored in sprinkler equipped cabinets. Some models have doors that close automatically in the event of fire. Glass bottles should stand in trays that will contain the liquid if the bottle breaks, thus limiting the spread of the contamination.

Other considerations in the storage of flammable liquids in the laboratory include ensuring that passageways and exits are not blocked, that accidental contact with strong oxidizing agents such as chromic(VI) acid, manganate(VII), chlorates(I, III, V, VII), or peroxides is not possible, and that sources of ignition are excluded (see also Section 7.1.2).

5.4.3. Toxic substances

Chemicals that have high chronic toxicity, including those classified as potential carcinogens, should be stored in ventilated storage areas in unbreakable, chemically resistant secondary containers. Only the quantity of toxic material that is required for the current experiment should be present in the working area. Storage vessels containing such substances should carry a label: *CAUTION: HIGH CHRONIC TOXICITY* or *CANCER SUSPECT AGENT* (Section 6.9). An inventory of these materials should be maintained. Additional safeguards may be required by law.

Storage areas for substances that have high acute or moderate chronic toxicity should exhibit a warning sign denoting the hazard. There should be limited access and adequate ventilation (Section 6.7, 6.9).

5.4.4. Compressed gases

Cylinders of compressed gases in the laboratory should be securely strapped or chained to a wall or bench-top to prevent their being knocked over accidentally. The safest practice is to have the cylinders chained to the outer wall of the laboratory building, with the gases piped into the laboratory. They should be capped when not in use, and kept away from sources of heat or ignition (Sections 5.2.6 and 8.1).

6

Procedures for Working with Substances that Pose Hazards Because of Acute Toxicity, Chronic Toxicity or Corrosivity

Many of the chemicals encountered in the laboratory are known to be toxic, corrosive, or both. New and untested substances that may be hazardous are also frequently encountered. Thus it is essential that all laboratory workers understand the types of toxicity, know the routes of hazardous exposure, and recognize the major classes of toxic and corrosive chemicals. During the planning of an experiment, the chemist should become familiar with the physical and chemical properties of the starting materials and expected products, and also with their potentially hazardous properties. Chemists are usually familiar with sources of physical and chemical data, but are less likely to know where to find information on toxicological properties. The bibliography at the end of this book includes sources of information on general types of hazards posed by chemicals in the laboratory, as well as on specific aspects, namely toxic hazards and toxicology, flammability, carcinogens, and chemical interactions.

In considering the toxicity hazards of an experiment, it is important to recognize that the combined toxic effect of two substances may sometimes be more than additive. For example, ethanol and carbon tetrachloride may both cause liver damage, but together the effect is more serious than would be expected from the simple addition of their individual actions. Because chemical reactions may contain mixtures of substances whose combined toxicities have not been evaluated, it is prudent to assume that mixtures of different substances will be more toxic than the most toxic ingredient contained in the mixture. Furthermore, chemical reactions involving two or more substances may form reaction products that are more toxic than the starting reactants. This possibility of generating toxic reaction products may not be anticipated by the laboratory worker when the reactants are mixed unintentionally. For example, inadvertent mixing of formaldehyde (a common tissue fixative) and hydrochloric acid could result in the generation of bis(chloromethyl) ether, a potent human carcinogen.

Exposure to hazardous chemicals can be prevented or minimized by following the procedures detailed elsewhere in this monograph. Only the most pertinent measures are stressed in this chapter.

6.1. ROUTES OF EXPOSURE

The main routes of exposure to chemicals are inhalation, ingestion, contact with skin and eyes, and "injection", i.e., by flying objects resulting from an explosion. Of these, inhalation and contact with skin and eyes present the greatest hazards in the laboratory.

6.1.1. Inhalation

Inhalation of toxic vapours, mists, gases, aerosols or dusts can produce poisoning by absorption through the mucous membrane of the mouth, throat, and lungs and can seriously damage these tissues by local action. Inhaled gases or vapours may pass into the capillaries of the lungs and be carried into the circulatory system. This absorption can be extremely rapid. The degree of injury depends on the toxicity of the material, its physico-chemical properties and its solubility in tissue fluids, as well as its concentration and the duration of exposure. Chemical reactivity and the time of response after exposure are not necessarily a measure of the degree of toxicity. Some chemicals such as mercury or benzene are cumulative poisons in the sense that they can produce body damage through exposure to small concentrations over a long period of time.

The threshold limit value (TLV) is a time-weighted average, air-borne concentration for a normal 8-hour workday to which nearly all workers may be repeatedly exposed without adverse effect. Many countries produce lists of TLVs or similar guides for chemicals commonly used in the laboratory. These are helpful in deciding whether special ventilation and other precautions need to be taken in the laboratory.

The most effective way to avoid exposure to toxic vapours, mists, gases, and dusts is to prevent the escape of such materials into the working atmosphere and to ensure adequate ventilation by the use of exhaust hoods (Section 3.2) and other forms of ventilation. Chemicals of unknown toxicity should not be smelled. If there is a special reason to do so, as in the course of identification, the container cap or a drop of substance on a piece of filter paper can be sniffed cautiously near a fume hood.

6.1.2. Ingestion

Many chemicals used in the laboratory are extremely dangerous if they enter the mouth and are swallowed (Section 2.7). The toxicities of different chemicals can be compared by acute or chronic toxicity testing in animals.

The acute toxicity of a chemical can be evaluated by determining the dose that is lethal to 50% of a group of tested animals of a given species when administered orally, or to the skin, or by some other route in a single dose (LD_{50}). This is expressed in grams or milligrams per kilogram of body weight. In sublethal doses, many chemicals may damage the tissues of the mouth, nose, throat, lungs, and gastrointestinal tract, and if absorbed through tissue, may produce systemic poisoning.

To prevent entry of toxic chemicals into the mouth, laboratory workers should wash their hands before handling food, smoking, or applying cosmetics, immediately after use of any toxic substance, and before leaving the laboratory. Food and drink should not be stored or consumed in areas where chemicals are being used, nor should cigarettes, cigars, and pipes be used in such areas. Chemicals should not be tasted, and the pipetting or siphoning of liquids should never be carried out by mouth.

6.1.3. Contact with skin and eyes

Contact with the skin is a frequent mode of chemical injury, a common result being localized irritation (Section 6.4). Many chemicals can be absorbed through the skin to produce systemic poisoning. The rate of absorption varies in different regions of the body, the main portals of entry being hair follicles, sebaceous glands, sweat glands, and cuts or abrasions of the outer layers of the skin. The numerous blood vessels in the follicles and glands facilitate the absorption of chemicals into the body.

Contact of chemicals with the eyes is of particular concern because these organs are so sensitive to irritants. Few substances are innocuous, and most are painful and irritating. A considerable number are capable of causing burns and loss of vision. Alkaline materials, phenols, or strong acids are particularly corrosive and can cause permanent loss of vision. The numerous blood vessels and the moisture of the eye effect rapid absorption of many chemicals.

Skin and eye contact with chemicals should be avoided by use of appropriate protective equipment (Sections 2.1-2.3). All persons in the laboratory should wear safety glasses. Face shields, safety goggles, bench shields and similar devices provide better protection for the eyes. Protection against skin contact may be obtained by use of gloves, laboratory coats, tongs, and other protective devices. Spills should be cleaned up promptly (Section 2.5.3).

In the event of skin contact, the affected areas should be flushed with water. Medical attention should be sought if symptoms persist. In the event of eye contact, the eye should be flushed with cold running water for 15 minutes and medical attention sought, whether or not symptoms persist (Section 2.7).

6.1.4. Injection

Exposure to toxic chemicals by injection can occur inadvertently through mechanical injury from sharp glass or metal objects contaminated with chemicals. Syringes which are used for hypodermic injections or in robotic apparatus are always potentially dangerous.

6.2. ACUTE AND CHRONIC TOXICITIES

The toxicity of a material results from its ability to damage or interfere with the metabolism of living tissue. An acutely toxic substance can cause damage as the result of a single exposure. A chronically toxic substance causes damage after repeated or long duration exposure. Hydrogen cyanide, hydrogen sulfide, and nitrogen(IV) oxides are examples of acute poisons. Chronic poisons including carcinogens and some metals and metal compounds, such as mercury, lead, and their derivatives, are particularly insidious because of their long latency periods, so that the cumulative effects of low exposure may not be apparent for many years. Some chemicals, e.g., bis(chloromethyl) ether, can have either high acute toxicity or high chronic toxicity, depending on the conditions of exposure. All new and untested chemicals should be regarded as toxic until proven otherwise.

6.3. POISONS WITH DEVELOPMENTAL TOXICITY

Some substances that are absorbed during pregnancy can cause adverse effects on the fetus. These effects include death of the fertilized egg, the embryo, or the fetus, malformations (teratologic effects), retarded growth, and postnatal functional defects.

Some of these substances have been demonstrated to be developmentally toxic in humans. These include organomercurials, lead compounds, and thalidomide, a drug that is no longer used as a sedative. Although maternal alcoholism is the leading known cause of developmental toxic effects in humans, there is no evidence that such effects can result from exposure to ethanol vapour under normal laboratory conditions. Many common substances have been shown to be developmentally toxic to animals at some exposure level, but usually this is an enormously higher level than is met in the course of normal laboratory work or any other normal activity. However, some substances do require special controls because of their potentially toxic properties. One example is formamide. Women of child bearing age should only handle this substance in a hood and should take precautions to avoid skin contact with the liquid because of the ease with which it can be absorbed. Cytotoxic anti-cancer drugs, should also, be included in this category.

As the period of greatest susceptibility is the first 8-12 weeks of pregnancy, a time when a woman may not know that she is pregnant, women of child bearing age should take care to minimize skin contact with laboratory chemicals. The following procedures are recommended:

1. The use of substances with developmental toxicity should be reviewed by the research supervisor, who will decide whether special procedures or warning signs are warranted. Consultation with appropriate safety personnel may be desirable. In cases of continued use of a known agent of this kind, the operation should be reviewed annually or whenever a change in operating conditions is made.
2. Substances with developmental toxicity that require special control should be stored in a ventilated area with limited access. The container should be labelled in a clear manner so as to emphasize the potential for developmental toxicity. If the storage container is breakable, it should be kept in an impermeable, unbreakable secondary container with sufficient capacity to retain the material if the primary container breaks. A log of the amount in storage should be kept.
3. Women of child bearing age should guard carefully against spills and splashes. Operations should be carried out in well ventilated areas. Appropriate safety apparel, especially gloves, should be worn.
4. Certain substances cause testicular toxicity in men. The same procedure for women of child bearing age should be followed.
5. Supervisors should be notified of all incidents of exposure or spills of these or other agents, that require special control. A qualified physician should be consulted concerning the acceptable level of exposure to women of child bearing age.

6.4. ALLERGENS

Chemical allergy is an adverse reaction to a chemical (an "allergen") that results from previous sensitization to that chemical through contact with it. The skin and lungs are particularly sensitive to allergic reactions, and a wide variety of substances have caused such reactions. Examples include relatively common substances as diazomethane, nickel, (di)chromate(VI), formaldehyde, isocyanates, and certain phenols. Because of this variety and the varying response of individuals, suitable gloves should be used whenever hand contact with products of unknown activity is probable (Section 2.2), and good ventilation is essential.

6.5. CORROSIVE CHEMICALS

The major classes of corrosive chemicals are strong acids and bases, dehydrating agents, and oxidizing agents. Some chemicals, e.g., concentrated sulfuric acid, belong to more than one class; it should not be forgotten that the strong organic acids are also corrosive. Inhalation of vapours or mists of these substances can cause severe bronchial irritation. These chemicals erode the skin and the respiratory epithelium and are particularly damaging to the eyes.

6.5.1. Strong acids

All concentrated strong acids can damage the skin and eyes. Exposed areas should be flushed promptly with water. Nitric(V), chromic(VI), and hydrofluoric acids are especially damaging because of the types of burns they inflict. Hydrofluoric acid, which produces slow-healing, painful burns, should be used only after thorough familiarization with recommended handling procedures (Section 6.8.2).

6.5.2. Strong bases

The common strong bases are potassium hydroxide, sodium hydroxide, and ammonia. Ammonia is a severe bronchial irritant and should be used only in a well-ventilated area. The metal hydroxides are extremely damaging to the eyes. If exposure occurs, the eyes should be washed immediately with cold running water for 15 minutes and an ophthalmologist be consulted to assess the need for further treatment.

6.5.3. Dehydrating agents

Strong dehydrating agents include concentrated sulfuric acid, sodium hydroxide, phosphorus(V) oxide, and calcium oxide. Because much heat is evolved on mixing these substances with water, mixing should always be carried out by continuously adding the agent to water so as to avoid violent reaction and spattering. These substances cause severe corrosive and thermal burns on contact with the skin because of their affinity for water. Affected areas should be washed promptly with large volumes of cold water.

6.5.4. Oxidizing agents

In addition to their corrosive properties, powerful oxidizing agents such as chromic(VI) acid and chloric(VII) acid present fire and explosion hazards on contact with most substances that can be oxidized. The hazards associated with the use of chloric(VII) acid are especially severe (Section 7.3); it should be handled only after thorough familiarization with recommended procedures. Strong oxidizing agents should be stored and used in glass or other inert containers (preferably break-resistant); corks or rubber stoppers should not be used. Reaction vessels containing significant quantities of these reagents should be heated in fibreglass mantles or sand baths rather than in oil baths.

6.6. TYPES OF HANDLING PROCEDURES

Recommendations for chemical handling procedures begin with the admonition that, even for substances of no known significant hazard, it is prudent to observe the good laboratory practices of Chapter 1, minimizing exposure by working in an exhaust hood, and wearing eye and hand protection and a laboratory coat or apron.

Many of the general recommendations for safe practices in the laboratory are especially applicable to the handling of corrosive substances. Here it is particularly important that attention be given to the use of protective apparel and safety equipment (Chapter 2). In addition, the storage, disposal, and clean-up of corrosive substances requires special care. Bottles of corrosive liquids should be stored in acid-resistant containers over polyethylene trays large enough to contain the contents of the bottle if it fractures. Most major suppliers provide acids in plastic-coated glass bottles, which are less likely to break than ordinary bottles. To ensure that mutually reactive chemicals cannot accidentally contact one another, such substances should not be stored on the same trays unless they are in unbreakable, corrosion-resistant secondary containers. In the disposal of corrosive substances, care must be taken not to mix them with other potentially reactive wastes. Spillage involving such substances should be contained, carefully diluted with water, and neutralized (Section 2.5.3).

To minimize the risk of handling toxic substances, two general procedures are recommended: *Procedure A* for substances of moderate chronic or high acute toxicity, and those whose toxicological properties are not known and *Procedure B* for substances of known high chronic toxicity. These procedures enable one to work safely with most substances, whatever their chemical, physical, or toxicological properties, whether known or unknown.

More restrictive procedures are sometimes required by law for specific highly toxic substances, notably carcinogens such as 2-naphthylamine and vinyl chloride. Laboratories must check whether there are national or local regulations that require special storage and working conditions, monitoring, record keeping, medical surveillance, or special disposal procedures for any chemicals that they use or might generate as by-products.

6.7. GENERAL PROCEDURES AND PRECAUTIONS FOR WORKING WITH SUBSTANCES OF MODERATE CHRONIC OR HIGH ACUTE TOXICITY (PROCEDURE A)

This procedure should be followed for laboratory operations with substances which, if used infrequently in small quantities, are not a significant carcinogenic hazard, but which could be so to people exposed to high concentrations or repeated small doses. Thus a substance that is not known to cause cancer in humans, but which has shown statistically significant, but low, carcinogenic potency in animals, generally should be handled according to this procedure. The procedure is also appropriate for substances with high acute toxicity, such as hydrogen cyanide, hydrogen sulfide, or diisopropyl fluorophosphate. It is likewise advisable to use it when dealing with chemicals for which little or nothing is known of their toxicological properties. Clearly some could be quite toxic.

Before setting up an experiment, the laboratory worker should be familiar with all of the potential hazards of substances that will be used.

The overall objective of Procedure A is to minimize exposure of the laboratory worker to toxic substances, by any route of exposure, through taking all reasonable precautions. Thus, the general precautions outlined in Chapter 1 should normally be followed whenever a toxic substance is being transferred from one container to another or is being subjected to some chemical or physical manipulation. It is particularly important to emphasize the following three precautions:

1. Protect hands and forearms by wearing either suitable long (gauntlet type) gloves (Section 2.2) and a laboratory coat to avoid contact of toxic material with the skin.
2. Procedures involving volatile toxic substances and those involving solid or liquid toxic substances that may generate aerosols should be conducted in a hood or other suitable containment system (Chapter 3). The hood should have been evaluated previously to establish that it is providing adequate ventilation and has an average face velocity of not less than 0.3 m/s.
3. After working with toxic materials, wash the hands and arms immediately. Never eat, drink, smoke, chew gum, apply cosmetics, take medicine, or store food in areas where toxic substances are being used.

These standard precautions will provide laboratory workers with good protection from most toxic substances. In addition, records that include amounts of material used and names of the workers involved should be kept as part of the laboratory notebook record of the experiment. To minimize hazards from accidental breakage of apparatus or spillage of toxic substances in the hood, containers should be stored in pans or on trays made of polyethylene or other chemically inert material, and apparatus should be mounted above pans or trays of the same type of material. Alternatively, the working surface of the hood can be fitted with a removable liner of absorbent plastic backed paper. Toxic vapours that are discharged from the apparatus should be condensed or absorbed in water or an aqueous solution to avoid adding toxic material to the hood exhaust air. Areas where toxic substances are being used and stored should have restricted access, and warning notices should be posted.

Toxic waste materials should be disposed of in the proper manner (Part III). The laboratory worker should be prepared for possible accidents or spills involving toxic substances. If a toxic substance comes into contact with the skin, the area should be washed well with water, or a safety shower should be used. If there is a major spillage outside the hood, the room or appropriate area should be evacuated, and measures taken to prevent exposure of other workers. Spillages should be cleaned up by personnel wearing suitable protective apparel (Section 2.3). If a spillage of a toxicologically significant quantity of harmful material occurs outside the hood, a supplied-air, full face respirator must be worn while cleaning it up. Contaminated clothing and shoes should be thoroughly decontaminated or incinerated.

The precautions to be taken with a chemical will vary with its properties. For example, methyl acrylate is volatile (b.p. 80 $^{\circ}$C), flammable, lachrymatory, and a skin irritant. Hence it should be handled well back in a hood, its container should be open for as little time as possible, and gloves of oil-resistant material (neoprene or nitrile rubber) should be worn. The user's front should be protected by a rubber apron if more than 1 gram is used. There should be no open flame in the vicinity. Strychnine, a high melting solid of low volatility, has a high acute toxicity. Consequently, prompt stoppering of an opened container would not be so important, and thin rubber gloves favouring manual dexterity would suffice, but draughts that might blow strychnine dust about must be avoided.

Detailed descriptions of how to handle diisopropyl fluorophosphate, hydrogen fluoride, and hydrogen cyanide, three well known laboratory chemicals with high acute toxicity, illustrate the use of Procedure A.

6.8. EXAMPLES OF THE USE OF PROCEDURE A
6.8.1. Diisopropyl fluorophosphate
Physical properties

Diisopropyl fluorophosphate (DFP) is a viscous liquid with a vapour pressure of 0.58 mm mercury at 20 $^{\circ}$C. It is soluble in water (1.54% w/w), but gradually decomposes to form hydrofluoric acid.

Toxicity

DFP was used as a chemical warfare nerve gas. Like the organophosphorus insecticides, it inactivates acetylcholinesterase, leading to the accumulation of acetylcholine. The toxic effects respond to treatment if the antidotes described below are used promptly.

DFP is absorbed by inhalation, ingestion, or topical contact. It causes pinpoint pupils as well as severe lachrymation and rhinitis. Other symptoms include weakness, wheezing, and tachycardia. Severe intoxication is indicated by ataxia, confusion, convulsions, and respiratory paralysis leading to death.

Handling procedures

DFP should be stored in ampoules containing the minimum amount necessary for use. The reagent is not expensive, and excess material should be neutralized and disposed of in a safe manner (Section 10.1.10). Ampoules containing 0.1, 0.4, and 1.0 g are available. These should be opened in a hood, using great care to avoid splattering, or in a glove box equipped with a high efficiency particulate air (HEPA) filter. Protective apparel should include disposable gloves and a face shield.

A large container of aqueous sodium hydroxide (2 mol/dm^3) should be available wherever DFP is being used so that the entire ampoule, if necessary, can be discarded into it. Decomposition of DFP is relatively rapid in alkaline solution. Once the ampoule has been opened, the required amount of DFP should be withdrawn with a disposable syringe and used immediately. The opened ampoule and gloves and other items that have come in contact with the DFP should be washed with the alkali before disposal. Any spillage should be neutralized with excess alkali.

Solutions of DFP should be handled only under conditions that provide adequate ventilation. Most cold rooms, for example, are not adequately ventilated. All solutions of DFP should be kept in tightly stoppered containers. Solutions should be made alkaline before disposal to ensure complete decomposition of any remaining DFP.

Emergency treatment

Treatment consists of immediate administration of atropine followed by the specific antidote *N*-methylpyridinium-2-aldoxime chloride (2-PAM, pralidoxime chloride), commercially available in many countries. Atropine blocks certain receptors upon which acetylcholine acts while pralidoxime chloride reactivates the covalently modified acetylcholinesterase. Both drugs should be on hand in any laboratory in which DFP is used, and medical assistance must be readily available. In some countries, 2-PAM is not available, in which case toxogonin, which has a similar effect, should be used.

The keys to safety in handling DFP are an appreciation of its toxicity, the presence of adequate ventilation, the use of alkali to treat spillage and all objects in contact with the reagent, and the availability at the laboratory of the two pharmaceuticals.

6.8.2. Hydrofluoric acid

Hydrofluoric acid (HF) vapour and dilute or concentrated solutions all cause severe burns. Inhalation of anhydrous HF or HF mists or vapours can cause severe respiratory tract irritation that may be fatal.

Physical properties

Anhydrous HF is a clear, colourless liquid that boils at 19.5 °C. As a result of its low boiling point and high vapour pressure, anhydrous HF must be stored in pressure containers. HF is commonly used as a 70% (w/v) aqueous solution. Hydrofluoric acid is miscible with water in all proportions and forms an azeotrope [38.3% HF (w/v)] that boils at 112 °C.

Toxicity

Anhydrous or concentrated aqueous HF causes immediate and serious burns to any part of the body. Dilute solutions and gaseous HF are likewise harmful, although several hours may pass before the acid penetrates the skin sufficiently to cause burning.

Wearing clothing or leather shoes and gloves that have absorbed small amounts of HF can result in serious delayed effects such as painful, slow-healing skin ulcers.

Hazards from fire or explosion

Hydrofluoric acid is non-flammable, but difficult to contain because it attacks glass, concrete, and some metals, especially cast iron and alloys that contain silica. It also attacks organic materials such as leather, natural rubber, and wood. Because aqueous HF can react with metallic containers and piping to form hydrogen, potential sources of ignition (sparks and flames) should be excluded from areas in which equipment containing the acid is in use.

Handling procedures

It is crucial to ensure adequate ventilation by working only in a hood so that safe levels (less than 3 cm^3/m^3) are maintained. All contact of the vapour or the liquid with eyes, skin, respiratory system, or digestive system must be avoided by using protective equipment such as face shields and neoprene or polyvinyl chloride gloves. The protective equipment should be washed after each use to remove all traces of HF. Safety showers and eyewash fountains should be available. Workers using HF must have received prior instruction concerning its hazards and be aware of correct protective measures and recommended procedure for treatment in the event of exposure.

Spills and leaks

The vapours of both anhydrous HF and aqueous 70% (w/v) HF produce visible fumes on contact with moist air. This characteristic can be useful in detecting leaks, but cannot be relied on because of atmospheric variability. Spillage of HF must be treated immediately in order to minimize the dangers of vapour inhalation, body contact, corrosion of equipment, and possible generation of dangerous gases. Spillage should be contained and diluted with water, and the resulting solution neutralized with lime (calcium oxide) before disposal.

Waste-disposal

Waste HF can be stirred gradually into a solution of excess slaked lime (calcium hydroxide) to precipitate the fluoride as calcium fluoride, which is insoluble and harmless (Section 10.2.3).

Emergency treatment

Any suspected contact with HF should be treated immediately by flushing the exposed area with large quantities of cold running water. Exposed clothing should be removed as quickly as possible while flushing. Medical attention should be obtained promptly, even if the injury appears slight. Meanwhile the burned area should be immersed in a mixture of cold or iced water. If immersion is impractical, a compress should be made by inserting ice cubes between layers of gauze.

If HF liquid or vapour has contacted the eyes, these should be flushed with large quantities of cold running water while the eyelids are held apart. This flushing should be continued for 15 minutes, and medical attention obtained promptly.

If HF vapour has been inhaled, the person should be removed immediately to an uncontaminated atmosphere and kept warm; medical help or advice should be obtained as soon as possible.

Ingestion of HF should be treated by drinking a large volume of cool water as quickly as possible. Do not induce vomiting. Again, medical help or advice should be sought as soon as possible.

6.8.3. Hydrogen cyanide

The procedures given below are appropriate for the safe use of hydrogen cyanide (HCN) and related compounds such as metallic cyanides and cyanogen halides, which may release the cyanide ion or cyanogen or generate HCN when acidified (e.g., sodium or potassium cyanide). "HCN" in these procedures is intended to be specific for hydrogen cyanide, but to indicate general applicability to precursor compounds. All users of HCN must study or be instructed in HCN handling procedures before starting to work with this material.

Physical properties

Hydrogen cyanide is a colourless liquid, b.p. 25.7 °C, flash point -17.8 °C (closed cup), explosive limits 6-41% (v/v). It is freely miscible with water.

Toxicity

Hydrogen cyanide is among the most toxic and rapidly acting of all poisonous substances. Exposure to high doses may be followed by almost instantaneous collapse, rapid cessation of respiration, and death. At lower dosages, the early symptoms include weakness, headache, confusion, nausea, and vomiting. In humans, the approximate fatal oral dose is 40 mg. Exposure to 3000 cm^3/m^3 of HCN in air is immediately fatal, while 200-480 cm^3/m^3 can be fatal after 30 minutes. Exposure to 18-36 cm^3/m^3 of HCN in air causes slight symptoms after several hours. The liquid is rapidly absorbed through the skin and mucous membranes.

Because of its low flash point and wide range of explosive mixtures, HCN presents a serious fire and explosion hazard. In addition, polymerization may occur from the action of the cyanide ion on HCN. This reaction is unpredictable, but once started, it may become violent and produce sharp increases in temperature and pressure. This may occur in pure unstabilized liquid HCN if a base is present. Hence amines, hydroxides, and cyanide salts that are capable of producing cyanide ions should not be added to liquid HCN without suitable precautions. If liquid HCN is heated above 115 °C in a sealed vessel, a violent exothermic reaction generally occurs.

Commercially available HCN is normally stabilized by the addition of a little phosphoric(V) acid, from which HCN can be removed by distillation. However, distilled HCN presents a greater explosion hazard than the stabilized material, so that the use of the distilled material should be avoided.

Handling procedures

All work with HCN must be confined to hoods, which should have a minimum face velocity of 0.3 m/s. Care must be exercised to prevent the contact of either liquid HCN or its vapour with the skin. Butyl rubber gloves should be worn at all times when working with HCN. Whenever work with HCN or related compounds is being carried out in a laboratory, there should be at least two people present in the area.

Signs warning that HCN is in use should be posted at each entrance to the laboratory area. *WARNING* or *NO ADMITTANCE* signs should be posted on the doors to fan lofts and roofs whenever cyanides are being used in hoods.

All equipment in which cyanides are used or produced should be placed in or over shallow pans or trays so that spillage or leakage will be contained.

In the event of spillage of HCN or cyanide solutions the contaminated area should be evacuated and any risk of exposure to individuals determined. If necessary, other parts of the building should be evacuated. It is usually better not to dilute or absorb spillage of volatile cyanides if they occur in well-ventilated areas. However, large spillages of solutions of cyanide salts, such as sodium cyanide or potassium cyanide, should not be allowed to remain because they can liberate HCN slowly over a long period. Such spillage should be diluted by personnel wearing appropriate air mask, hand and foot protection, and then carefully treated with chlorate (I) (household bleach, 5.25% NaOCl w/v) to oxidize the cyanide to cyanate (Section 10.2.3). After the oxidation, the spill can be absorbed and disposed of in the proper way.

Detection

Hydrogen cyanide has a characteristic odour that resembles that of bitter almonds. However, many people cannot smell it in low concentrations. The ability to smell HCN at all is a genetically determined characteristic, and some people cannot smell it even in relatively high concentration. This group includes approximately 15% of the Caucasian and Japanese populations. Accordingly, odour should not be relied on. Vapour-detector tubes sensitive to $1 \text{ cm}^3/\text{m}^3$ of HCN are available commercially. The presence of free cyanide ion in aqueous solution can be detected by treating an alkaline aliquot of the sample with iron(II) sulfate and an excess of sulfuric acid. A precipitate of Prussian blue indicates that free cyanide ion is present.

Cyanide salts should not be stored or transported together with acids. An open bottle of NaCN can generate HCN in humid air.

Storage and dispensing

Storage of liquid HCN (except in commercial cylinders) in laboratory areas should be prohibited without special permission. If such storage is necessary, it must be in an exhaust hood or an area that has independent ventilation facilities. Signs should be posted restricting access to the area of the ventilation exhaust. Liquid HCN should be dispensed from cylinders only by trained personnel.

Waste-disposal

Hydrogen cyanide and waste solutions containing cyanides must not be emptied into the sewer or left to evaporate in an exhaust hood. Such materials can be disposed of by incineration in ethanol solution or, in small quantities, by oxidation to cyanate with chlorate(I) (Section 10.2.3).

Emergency treatment

An HCN first aid kit, including an oxygen cylinder equipped with pressure gauge and needle valve, should be available on each floor of a building in which work with cyanides is in progress. First aid kits and

102   *Chemical Safety Matters*

oxygen cylinders may be located near the work area, but they should not be in the same room. The main cylinder valve should be kept closed except when the cylinder is being used or checked. A tag should be attached to the oxygen cylinder indicating that it is reserved for emergency HCN first aid.

The HCN first aid kit should contain a facepiece and length of rubber tubing for administering oxygen, a 1 dm^3 bottle of 1% w/v sodium thiosulfate solution and a box of amyl nitrite pearls.

Any person exposed to HCN should be removed from the contaminated atmosphere immediately, taking care that those involved in the rescue are not exposed to HCN vapour. Any contaminated clothing should be removed and the affected area deluged with water. The person should be kept warm and breathing. An amyl nitrite pearl should be held under the nose for not more than 15 s/min (excess nitrite will reduce the blood pressure), and oxygen should be administered in the intervals. If the person is not breathing, artificial resuscitation should be commenced. When breathing starts, amyl nitrite and oxygen should be administered immediately. If HCN or one of its salts has been ingested, the person should be given 500 cm^3 or more of cool water, and vomiting should be induced promptly. The person must receive medical care immediately.

6.9. ADDITIONAL PROCEDURES AND PRECAUTIONS FOR WORKING WITH SUBSTANCES OF KNOWN HIGH CHRONIC TOXICITY (PROCEDURE B)

This procedure should be followed with those substances believed to be moderately to highly toxic, even when they are used in small amounts. A substance that has caused cancer in humans or has shown high carcinogenic potency in test animals will generally require the use of this procedure. However, other factors such as the physical form, volatility, the potential kind and duration of exposure, and the quantity of substance to be used, should also be taken into account.

A substance is deemed to have moderate to high carcinogenic potency in test animals if it causes a statistically significant tumour incidence: after inhalation exposure of 6-7 hours per day, 5 days per week, for a significant portion of a lifetime to dosages of less than 10 mg/m^3 in air; after repeated skin application of less than 20 mg/kg of body weight per week; after oral dosages of less than 50 mg/kg of body weight per day.

All of the procedures and precautions described under Procedure A (Section 6.8) should be followed when working with substances known to have high chronic toxicity. In addition, the precautions described below should be used. Each laboratory worker's plans for experimental work and for disposing of waste materials should be approved by the laboratory supervisor. Consultation with the departmental safety officer should be sought to ensure that the toxic material is contained effectively during the experiment and that waste materials are disposed of in a safe manner. Substances in this high-chronic-toxicity category include certain metal compounds (e.g., dimethylmercury and nickel carbonyl) and compounds classified as potent carcinogens, for exmples: benzo[a]pyrene; 3-methylcholanthrene; 7,12-dimethylbenz[a]anthracene; dimethylcarbamoyl chloride; hexamethylphosphoramide; 2-nitronaphthalene; 1,3-propane sultone (1-oxa-2-thiacyclopentane 2,2-dioxide); many N-nitrosamines; many N-nitrosamides; bis(chloromethyl) ether; aflatoxin B1; 2-naphthylamine; and 2-acetylaminofluorene.

An accurate record of the amounts of such substances being stored and used, dates of use, and names of users must be maintained.

It may be appropriate to keep such records as part of the account of experimental work in the laboratory workers' research notebooks, but it must be understood that the research supervisor is responsible for ensuring that accurate records are kept. Any volatile substances with high chronic toxicity should be stored in a ventilated storage area in a secondary tray or container that has sufficient capacity to contain the material and cannot react with the substance should the primary container accidentally break or spill. All containers of substances in this category should have labels that identify the contents and include a warning such as: *WARNING! HIGH CHRONIC TOXICITY* or *CANCER-SUSPECT AGENT.* Storage areas for substances in this category should have restricted access, and special signs should be displayed. Any area used for storage of substances of high chronic toxicity should be maintained under negative pressure with respect to surrounding areas.

All experiments with, and transfers of, such substances should be carried out in a controlled area. A controlled area is defined as a laboratory, or a portion of a laboratory, or a facility such as an exhaust hood or a glove box that is designated for the use of highly toxic substances. Its use need not be restricted to the handling of toxic substances if all personnel who have access to the controlled area are aware of the nature of the substances being used and the precautions that are necessary. When a negative-pressure glove box is used in which work is carried out through attached gloves, the ventilation rate in the glove box should be at least two volume changes per hour, the pressure should be at least 13 mm of water lower than that of the external environment, and the exit gases should be passed through a trap or efficient filter. If positive-pressure glove boxes are used to provide an inert, anhydrous atmosphere for working with highly toxic compounds, the box should be thoroughly checked for leaks, and the exit gases passed through a suitable trap or filter. Glove boxes should be regarded as permanently contaminated areas where the contamination can be transferred through the gloves. It is advisable, therefore, to wear additional protective gloves.

Laboratory vacuum pumps used with substances of high chronic toxicity should be protected by high efficiency scrubbers or filters and vented into an exhaust hood. The design of a particular vacuum pump should be taken into consideration if regular decontamination is likely. The decontamination of a vacuum pump should be carried out in an exhaust hood. Controlled areas should be clearly marked with a conspicuous sign such as: *WARNING: TOXIC SUBSTANCE IN USE* or *CANCER-SUSPECT AGENT: AUTHORIZED PERSONNEL ONLY*. Only authorized and instructed personnel should be allowed access to controlled areas.

Proper gloves (Section 2.3) should be worn when handling substances that have high chronic toxicity. The laboratory worker or the research supervisor may deem it advisable to use other protective apparel (Chapter 2), such as a low permeability apron covered by a disposable coat. Surfaces on which high chronic toxicity substances are handled should be protected from contamination by using chemically resistant trays that can be decontaminated after the experiment or by using dry, absorbent, plastic-backed paper that can be disposed of after use.

On leaving a controlled area, laboratory workers should remove protective apparel and thoroughly wash the hands, forearms, face, and neck. If disposable apparel or absorbent paper liners have been used, these items should be placed in a closed impervious container that should then be labelled in a clear manner such as: *CAUTION: CONTENTS CONTAMINATED WITH SUBSTANCES OF HIGH CHRONIC TOXICITY*. Non-disposable protective apparel should be thoroughly washed, and containers of disposable apparel and paper liners should be incinerated.

Wastes and other contaminated materials should be collected together with the washings from vessels. These should be decontaminated chemically or placed in closed, suitably labelled containers for incineration away from the controlled area. If chemical decontamination is to be used, a method should be chosen that will convert all of the toxic materials into non-toxic materials. For example, residues and wastes from experiments in which β-propiolactone or bis(chloromethyl) ether have been used should be treated with aqueous or ethanolic alkali (Sections 10.1.2, 10.1.10) as appropriate.

In the event that chemical decontamination is not feasible, wastes and residues should be placed in an impervious container that should be closed and labelled appropriately, for example: *CAUTION: COMPOUNDS OF HIGH CHRONIC TOXICITY* or *CAUTION: CANCER-SUSPECT AGENT*. Transfer of contaminated wastes from the controlled area to the incinerator should be carried out under the supervision of authorized personnel. Generally liquid wastes should be placed in suitable bottles (preferably polyethylene), and these should be transported in closed plastic or metal pails of sufficient capacity to contain the material in case of accidental breakage of the primary container.

Normal laboratory work should not be resumed in a space that has been used as a controlled area until it has been decontaminated. Work surfaces should be thoroughly washed and rinsed. If experiments have involved the use of finely divided solid materials, surfaces should be cleaned by wet mopping or by use of a vacuum cleaner equipped with an HEPA filter. Dry sweeping should not be carried out. All equipment known or suspected to have been in contact with substances of high chronic toxicity should be washed and rinsed before being removed from the controlled area.

In the event of continued experimentation with a substance of high chronic toxicity (e.g., if a worker regularly uses toxicologically significant quantities of such a substance three times a week), a qualified physician should be consulted to determine whether it is advisable to establish a regular schedule of medical surveillance or biological monitoring.

The use of Procedure B is illustrated by the following descriptions for benzo[*a*]pyrene, mercury and its compounds, and *N*-nitrosodiethylamine.

6.10. EXAMPLES OF THE USE OF PROCEDURE B

The procedures and precautions described in Procedure A must all be practised.

6.10.1. Benzo[*a*]pyrene (3,4-benzpyrene)

Although the information given below applies specifically to benzo[*a*]pyrene, the general precautions and procedures are also applicable to other carcinogenic polycyclic aromatic hydrocarbons.

Physical and chemical properties

Benzo[*a*]pyrene, $C_{20}H_{12}$, is a pale yellow crystalline solid that melts at 176-177 $^{\circ}$C and boils at 310-312 $^{\circ}$C/10 mm. It is readily soluble in benzene, xylene, and acetone, sparingly soluble in ethanol and methanol, and almost insoluble (0.005 mg/dm^3 at 27 $^{\circ}$C) in water.

Toxicity

Benzo[*a*]pyrene is a potent carcinogen that has produced tumours in all nine species for which experimental tests have been reported. Both local (skin and subcutaneous tissues) and systemic (lung and liver) carcinogenic effects have been observed. Tumours have resulted from skin application, oral administration, and parenteral injection. Benzo[*a*]pyrene is an example of a compound that is mutagenic in systems that possess mixed function oxidase systems for metabolism.

Handling procedures

All work with benzo[*a*]pyrene in quantities in excess of a few milligrams should be carried out in a well ventilated hood or in a glove box equipped with an HEPA filter. All work should be carried out in apparatus that is contained in or mounted above unbreakable pans that will contain any spillage. If very small amounts are being used, a disposable mat may be adequate to contain possible spillage. All containers should bear a label such as: *CANCER-SUSPECT AGENT*. All personnel who handle benzo[*a*]pyrene should wear plastic or latex gloves and a fully buttoned laboratory coat.

Storage and use

Bottles of benzo[*a*]pyrene should be stored and transported in unbreakable outer containers.

Clean-up of spillages and waste-disposal

Disposal of benzo[*a*]pyrene is best carried out by oxidation, e.g., by high temperature incineration or, for gram quantities, by the use of potassium manganate(VII) in sulfuric acid (Section 10.1.1). Prior to incineration of liquid wastes, solutions should be neutralized if necessary, filtered to remove solids, and put in a polyethylene container for transport. All equipment should be thoroughly rinsed with solvent for decontamination and this solvent should be added to the wastes to be incinerated. Great care should be taken to prevent contamination of the outside of the solvent container.

Solid reaction wastes should be incinerated or decomposed by chemical oxidation. Alternatively, solid reaction wastes can be extracted with solvent which is added to other liquid waste for incineration. Any contaminated rags and paper should be incinerated. Contaminated solid materials should be enclosed in sealed plastic bags that are labelled *CANCER-SUSPECT AGENT* and contain the name and quantity of the carcinogen. The bags should be stored in a well-ventilated area until they are incinerated.

6.10.2. Mercury and its compounds

Metallic mercury is widely used in laboratory instruments, and mercury compounds are used in many laboratory peocedures.

Physical properties

Mercury (relative atomic mass 200.6) is a dense, silvery white, shining metal that is liquid at room temperature. It melts at -38.9 $^\circ$C, boils at 356.6 $^\circ$C, and has a relative density of 13.594 at 4 $^\circ$C. It is insoluble in water. Its vapour is colourless, odourless, and tasteless.

Toxicity

Metallic mercury and mercury compounds can be absorbed into the body by inhalation, ingestion, or contact with the skin. Mercury is a subtle poison, the effects of which are cumulative and not readily reversible. The TLV for mercury is 0.05 mg/m^3; for its inorganic and aryl derivatives, 0.10 mg/m^3; and for its alkyl derivatives, e.g., dimethylmercury or ethylmercuric chloride, 0.01 mg/m^3.

Mercury poisoning by chronic inhalation can cause emotional disturbances, unsteadiness, inflammation of the mouth and gums, general fatigue, memory loss, and headaches. Kidney damage may result from poisoning by mercury salts.

In most cases of exposure by chronic inhalation, the symptoms of poisoning gradually disappear when the source of exposure is removed. However, improvement may be slow and complete recovery may take years. Skin contact with mercury compounds produces irritation. Soluble mercury salts can produce poisoning by absorption through the intact skin.

Handling procedures

Every effort should be made to prevent spillage of metallic mercury because the substance is extremely difficult to retrieve. Droplets get into cracks and crevices, under table legs, and under and into equipment. If spillage is frequent, each small source of mercury vapour may eventually contribute to raising its concentration in the laboratory air above the allowable limit.

Storage

Containers of mercury should be kept closed and stored in secondary containers in a well ventilated area. Instruments or apparatus containing mercury should be placed in an enamel or plastic tray that can be cleaned easily and is large enough to contain the mercury, as a precaution against breakage. Transfers of mercury from one container to another should be carried out in a hood and over a tray to confine any spillage.

Clean-up of spillage

Pools and droplets of metallic mercury can be made to coalesce and be collected as described in Section 9.1.1. A mercury vapour analyser should be available for determining the effectiveness of the clean-up operation.

If mercury has been spilled on the floor, the workers involved in the clean-up and decontamination activities should wear plastic shoe covers. When the clean-up is complete, the shoe covers should be disposed of and the workers should thoroughly wash their hands, arms, and face several times.

Spilled mercury compounds or solutions can be cleaned up by any method that does not cause excessive airborne contamination or skin contact.

Waste-disposal

Significant quantities of metallic mercury from spillage or broken thermometers or other equipment, and contaminated mercury from laboratory activities, should be collected in thick-walled high-density polyethylene bottles for reclamation (Section 9.1.1).

Rags, sponges, and shoe covers used in clean-up activities, and broken thermometers containing small amounts of residual mercury, should be placed in a sealed plastic bag, labelled, and disposed of by burial in a lab pack.

6.10.3. *N*-Nitrosodiethylamine (diethylnitrosamine)

Although the information given below applies specifically to *N*-nitrosodiethylamine [$(C_2H_5)_2N\text{-}NO$], the general precautions and procedures are also applicable to dimethylnitrosamines and other dialkylnitrosamines. It should be noted that the use of *N*-nitrosodimethylamine is regulated by some governments when it is present at concentrations greater than 1% in a solution or mixture.

Physical and chemical properties

N-Nitrosodiethylamine is a volatile yellow liquid that has a boiling point of 177 °C and a density (d_4^{20}) of 0.9422 g/cm^3. It is lipid soluble, dissolves in many organic solvents, and also has a water solubility of approximately 100 g/dm^3. It is stable for several days at room temperature in neutral or alkaline aqueous solutions, but less stable in strong acid. *N*-Nitrosodiethylamine can be oxidized to the corresponding nitramine or reduced to the hydrazine or amine.

Toxicity

N-Nitrosodiethylamine is a potent liver poison that can cause death from hepatic insufficiency in experimental animals. It is carcinogenic in at least 10 animal species, including subprimates. The main targets for its carcinogenic activity are the liver, lung, oesophagus, trachea, and nasal cavity. Although extensive data are not available on the toxicity of *N*-nitrosodiethylamine in humans, the closely related compound *N*-nitrosodimethylamine has caused extensive liver damage as a consequence of industrial exposure. Ingestion, inhalation, and skin application have all caused serious toxic effects in animals.

Handling procedures

All work with *N*-nitrosodiethylamine should be carried out in a well-ventilated hood or in a glove box equipped with an HEPA filter. All vessels containing *N*-nitrosodiethylamine should be kept closed. Work should be carried out in apparatus that is contained in, or mounted above, stainless steel pans that will contain any spillage. All containers should bear a label such as: *CANCER-SUSPECT AGENT*. All personnel who handle the material should wear plastic, latex, or neoprene gloves and a fully buttoned laboratory coat.

Storage

All bottles of *N*-nitrosodiethylamine should be stored and transported within an unbreakable outer container. Storage should be in a ventilated storage cabinet, or a hood.

Clean-up of spillages and waste-disposal

N-nitrosodiethylamine is chemically stable under normal conditions. Disposal is therefore best carried out by incineration or reduction to the amine (Section 10.1.15). Solutions to be incinerated should be neutralized if necessary, filtered to remove solids, and put in closed polyethylene containers for transport. All equipment should be thoroughly rinsed with solvent, which should then be added to the liquid waste for incineration. Great care should be taken to prevent contamination of the outside of the solvent container. If possible, solid wastes should be incinerated, otherwise the *N*-nitrosodiethylamine should be extracted and the extracts added to the liquid waste. Similarly, any contaminated rags or paper should be incinerated. Contaminated solid materials should be enclosed in sealed plastic bags that are labelled *CANCER-SUSPECT AGENT* and with the name and amount of the carcinogen. The bags should be stored in a well-ventilated area until they are incinerated.

Spillage of *N*-nitrosodiethylamine can be absorbed by Celite®, which should then be carefully collected (avoid dusts; do not sweep), and the surface thoroughly cleaned with a strong detergent solution, but no detergent is capable of degrading this nitroso compound. If a major spillage occurs outside a ventilated area, the room should be evacuated and the clean-up operation carried out by persons equipped with self-contained respirators. Those involved in this operation should wear rubber gloves, laboratory coats, and plastic aprons or equivalent protective apparel.

6.11. SPECIAL PROCEDURES AND PRECAUTIONS FOR ANIMAL STUDIES OF SUBSTANCES WITH HIGH CHRONIC TOXICITY

Animal studies of substances with high chronic toxicity can present a special exposure hazard to the experimenter if aerosols or dusts containing the toxic substance are formed. They can become dispersed throughout the laboratory or animal quarters by way of the animal food, urine, or faeces. Accordingly, procedures to minimize such formation should be implemented.

Administration of experimental substances by injection or gavage minimizes contamination of the laboratory. If toxic substances are administered in the diet, a relatively closed caging system should be adopted, either with the cages under negative-pressure or with a horizontal laminar airflow that is directed toward HEPA filters. In removing contaminated bedding or cage matting, a vacuum cleaner equipped with an HEPA filter can be used or the bedding can be wetted to reduce dusts. All diets containing toxic substances should be mixed in closed containers in a hood. Those carrying out such operations should wear plastic or rubber gloves and a fully buttoned laboratory coat or equivalent clothing. If exposure to aerosols cannot be controlled by other means, a respirator should be used (Section 2.4.2).

When large-scale studies are being carried out with highly toxic substances, special facilities or rooms with restricted access are preferable. If the caging system for the test animals does not adequately protect the personnel, a jumpsuit and shoe and head coverings should be worn.

7

Procedures for Working with Substances that Pose Flammable or Explosive Hazards

Flammable substances are among the most common of the hazardous materials found in laboratories. However, the ability to vaporize and ignite, and to burn or explode varies with the specific class of substance. Prevention of fires and explosions requires a knowledge of the flammability characteristics (limits of flammability, ignition requirements, and burning rates) of combustible materials likely to be encountered under various conditions of use (or misuse), and of the appropriate procedures for handling such substances.

7.1. FLAMMABILITY AND EXPLOSIVITY OF MIXTURES OF AIR WITH GASES, LIQUIDS AND DUSTS

7.1.1. Properties of flammable substances

Flammable substances are those that readily catch fire and burn in air. A flammable liquid does not itself burn; it is the vapour evaporated from the liquid that burns. The rate at which different liquids produce flammable vapours depends on their vapour pressure, which increases with temperature. The degree of fire hazard depends also on the ability to form a combustible or explosive mixture with air, the ease of ignition of mixtures, and the relative density of the liquid with respect to water, and of the vapour with respect to air.

Flash point

An open beaker of diethyl ether placed on the bench next to a Bunsen burner will ignite, whereas a similar beaker of diethyl phthalate will not. The difference in behaviour is due to the fact that the ether has a much lower flash point. The flash point is the lowest temperature, as determined by standard tests, at which a liquid gives off vapour in sufficient concentration to form an ignitable mixture with air near the surface of the liquid within the test vessel. Many common laboratory solvents and chemicals have flash points that are below room temperature.

Ignition temperature

The ignition temperature (auto-ignition temperature) of a substance is the minimum temperature required to initiate or cause self-sustained combustion independent of the heat source. A steam line or a glowing light bulb may ignite carbon disulfide (ignition temperature 80 ^{o}C). Diethyl ether (ignition temperature 160 ^{o}C) can be ignited by the surface of a hot-plate.

Limits of flammability

It is possible for a flammable liquid to be above its flash point and yet not ignite in the presence of an adequate energy source because the fuel-air mixture is too lean or too rich for combustion to occur. Each flammable gas and liquid (as a vapour) has two limits defining the range of concentrations in air that will ignite and may explode.

The lower flammable limit or lower explosive limit (LEL) is the minimum concentration (percent by volume) of the vapour in air below which a flame is not propagated when an ignition source is present. Below this concentration, the mixture is too lean to burn. The upper flammable limit or upper explosive limit (UEL) is the maximum concentration (percent by volume) of the vapour in air above which a flame is not propagated. Above this concentration, the mixture is too rich to burn. The flammable range (explosive range) comprises all concentrations between the lower and the upper explosive limits. This range becomes wider with increasing temperature.

With extremely flammable substances, notably diethyl ether and carbon disulfide, the upper limit of the flammability range provides little margin of safety. When such a solvent is spilled in the presence of an energy source, the LEL is reached very quickly and a fire or explosion will ensue before the UEL can be reached.

Spontaneous ignition

Spontaneous ignition or combustion takes place when a substance reaches its ignition temperature without the application of external heat. The possibility of spontaneous combustion should be considered, especially when materials are stored or disposed of. Materials susceptible to spontaneous combustion include oily rags, dust accumulations, organic materials mixed with strong oxidizing agents [such as nitric(V) acid, chlorates(V), manganates(VII), peroxides, and sulfates(VII)], alkali metals such as sodium and potassium, finely divided pyrophoric metals, and phosphorus.

The flash points, boiling points, flammable limits, and ignition temperatures of several common laboratory chemicals are given in Table 7.1. It should be remembered that tabulations of properties of flammable substances are based on standard tests for which the conditions may be very different from those encountered in practical use. Large safety limits should therefore be used when working with such compounds. For example, the published flammable limits for vapours assume a uniform mixture with air. In practice, local concentrations may be much higher than the overall composition of the mixture suggests. Thus it is good practice to set the maximum allowable concentration for safe working conditions significantly below tabulated LEL values. A value of 20% below the LEL is generally recognized as safe.

7.1.2. Handling flammable liquids

Among the most hazardous liquids are those that have flash points at or below room temperature, particularly if their range of flammability is broad. Materials with flash points above the maximum ambient summer temperature do not usually form ignitable mixtures with air under normal (unheated) conditions. However, as shown in Table 7.1, many commonly used substances are potentially hazardous, even under relatively cool conditions.

Sources of ignition

For a fire to occur, three conditions must exist simultaneously: a concentration of flammable gas or vapour that is within the flammable limits of the substance, an oxidizing atmosphere, usually air, and a source of ignition. Absence of any of these three will prevent the start of a fire. In most situations, air cannot be excluded. Fire is prevented, therefore, by exclusion of the co-existence of flammable vapours and an ignition source. Spillage of flammable liquids is always a hazard; thus strict control of ignition sources is mandatory.

Electrical equipment (Chapter 4), open flames, static electricity, burning tobacco, lighted matches, hot surfaces and other similar sources can cause ignition of flammable substances. When these materials are used in the laboratory, close attention should be given to all potential sources of ignition in the vicinity. The vapour from all flammable liquids is heavier than air and capable of travelling considerable distances. This possibility should be recognized, and special note taken of ignition sources that are beneath the level at which such substances are being used.

Flammable vapours from spilled chemicals have been known to descend into stairwells and elevator shafts and ignite at a lower storey. If the path of the vapour within the flammable range is continuous, the flame will propagate back to its source.

Table 7.1. Flash points, boiling points, ignition temperatures, and flammable limits of some common laboratory chemicals

Chemical	Flash point (°C)	Boiling point (°C)	Ignition temperature (°C)	Flammable Limit (percent by volume in air) Lower	Upper
Acetaldehyde	-37.8	21.1	175	4.0	60.0
Acetone	-17.8	56.7	465	2.6	12.8
Benzene	-11.1	80.0	560	1.3	7.1
Carbon disulfide	-30.0	46.1	80	1.3	50.0
Cyclohexane	-20.0	81.7	245	1.3	8.0
Diethyl ether	-45.0	35.0	160	1.9	36.0
Ethanol	12.8	78.3	365	3.3	19.0
n-Heptane	-3.9	98.3	215	1.0	6.7
n-Hexane	-21.7	68.9	225	1.1	7.5
Isopropyl alcohol	11.7	82.8	399	2.0	12.0
Methanol	11.1	64.9	385	6.7	36.0
Methyl ethyl ketone	-6.1	80.0	515	1.8	10.0
Pentane	-40.0	36.1	260	1.5	7.8
Styrene	32.2	146.1	490	1.1	6.1
Toluene	4.4	110.6	480	1.2	7.1
p-Xylene	27.2	138.3	530	1.1	7.0

Metal lines and large metal vessels such as drums and pilot-plant vessels from which large quantities of flammable substances are discharged should be electrically grounded (Figure 15) to prevent a static spark between the exiting liquid and the metal (Section 5.2.3). If either of the containers is non-metallic (especially plastic), the connection to ground should be made with the liquid, and the flow rate should be kept as low as possible (Figure 16).

Use of flammable substances

The safe handling of flammable materials necessitates the following basic precautions:

1. Flammable substances should be handled only in areas that are free of ignition sources.
2. Flammable substances must never be heated with an open flame. Alternative heat sources include steam baths, water baths, oil baths, heating mantles, and hot-air baths (Section 4.6).

Fig. 15 Bulk storage of a flammable liquid in a metal drum maintained at ground (earth) potential to prevent static sparking.

3. When transferring flammable liquids in or from metal equipment, static-generated sparks should be avoided by using grounding straps (Section 5.2.3).
4. Ventilation is one of the most effective ways to prevent the formation of flammable mixtures. An exhaust hood should be used whenever appreciable quantities of flammable substances are transferred from one container to another, allowed to stand in open containers, or heated in open containers.

7.1.3. Flammable or explosive gases and liquefied gases

Compressed or liquefied gases present hazards in the event of fire because heat will cause the pressure in the container to increase and may rupture it. Leakage or escape of flammable gases can produce an explosive atmosphere in the laboratory. Flammable gases in common use in laboratories include acetylene, hydrogen, ammonia, hydrogen sulfide, and carbon monoxide. Acetylene and hydrogen have very wide flammability limits, which add to their potential fire and explosion hazard.

Even if a substance is not under pressure, it is more concentrated in the form of a liquefied gas than in the vapour phase and may evaporate rapidly. Oxygen, in particular, is an extreme hazard; liquid air is almost as dangerous, because if it is allowed to boil freely, it will develop an increasing concentration of oxygen (b.p. -183 $^{\circ}$C) because nitrogen (b.p. -196 $^{\circ}$C) will

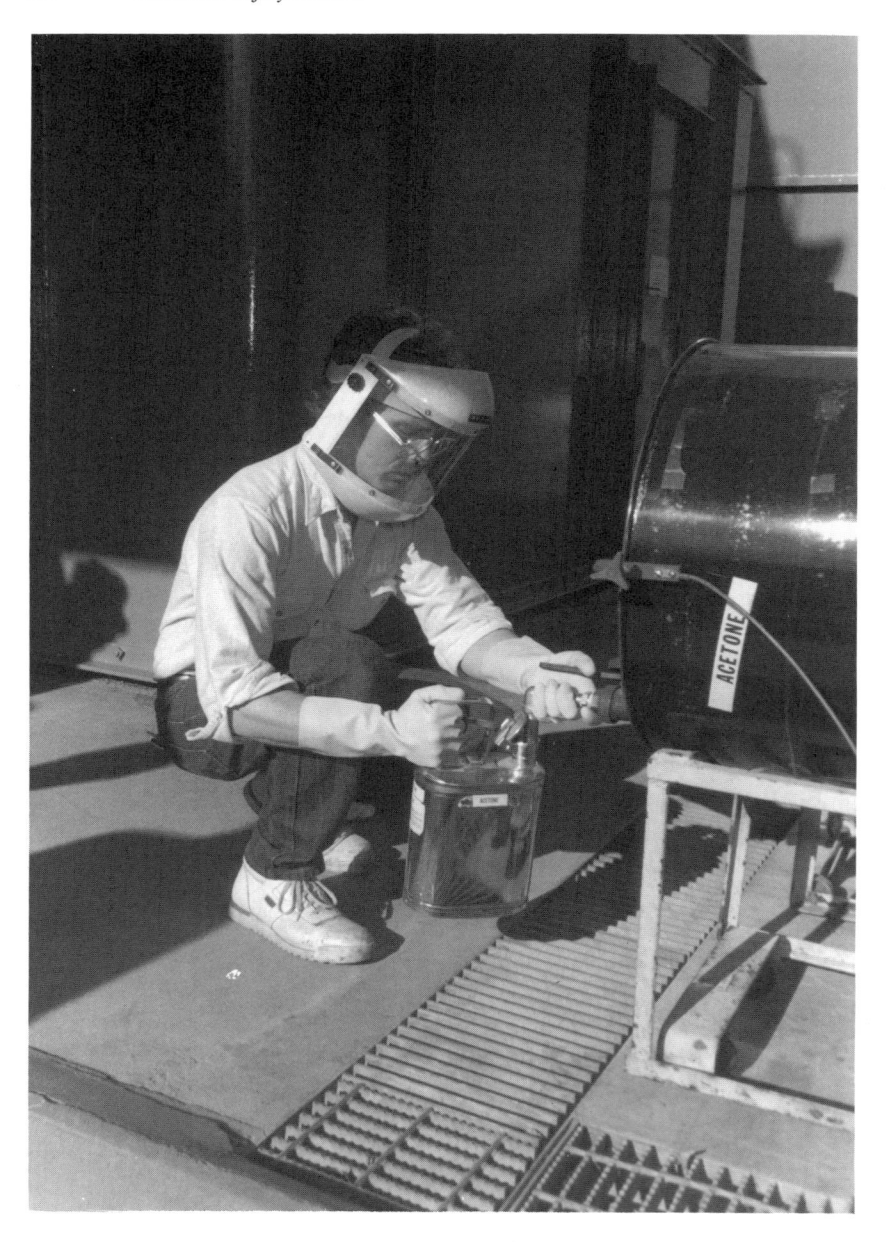

Fig. 16 In drawing off a flammable liquid from bulk storage, the metal receiving vessel is kept in contact with the spigot which is at ground (earth) potential.

boil away first. Even liquid nitrogen, if it has been standing open for some time, will have condensed enough oxygen from the air to require careful handling. When a liquefied gas is used in a closed-system, pressure may build up, so that a system to relieve pressure is required. If the liquid is flammable, explosive concentrations may develop.

7.1.4. Dusts

Air suspensions of oxidizable particles such as magnesium powder, zinc dust, or flowers of sulfur constitute powerful explosive mixtures. Care must be taken to avoid exposure to ignition sources, particularly static discharges.

7.2. HIGHLY REACTIVE AND INTRINSICALLY EXPLOSIVE CHEMICALS

An important general principle is that chemical reaction rates almost always increase rapidly as the temperature increases. If the heat evolved in a reaction is not dissipated, the reaction rate may increase until it becomes uncontrollable. This fact is particularly important when scaling-up a reaction, and sufficient cooling and heat exchange surface must be provided in the design of the apparatus to keep the reaction under control.

Some chemicals decompose when heated. Slow decomposition may not be noticed on a small-scale, but on a large-scale, or if the evolved heat and gases are confined, an explosive situation may develop. The heat-initiated decomposition of some substances, such as certain peroxides, can be almost instantaneous. Furthermore, the hazard is associated, not just with the total energy released, but with the extraordinarily high rate of a detonation reaction. The fast explosion of even milligram quantities can drive small fragments of glass or other matter deep into the body.

Light, mechanical shock, certain catalysts and even impurities may initiate explosive reactions. Thus, light can cause a mixture of hydrogen and chlorine to react explosively. Examples of shock-sensitive materials are acetylides, azides, and peroxides. Acids, bases, and other substances catalyse the explosive polymerization of acrolein, and many metal ions can catalyse the violent decomposition of hydrogen peroxide.

Not all explosions result from chemical reactions. A dangerous explosion with a physical basis can occur if a hot liquid, such as oil, is brought into sudden contact with one of lower boiling point, such as water. The rapid vaporization of the lower-boiling substance can be hazardous to personnel and destructive to equipment.

7.2.1. Organic peroxides

This group of intrinsically explosive chemicals includes dialkyl peroxides (e.g., $t\text{-}C_4H_9OOC_4H_9\text{-}t$), diacyl peroxides (e.g.,

$C_6H_5C(O)OOC(O)C_6H_5$), and hydroperoxides (e.g., $t\text{-}C_4H_9OOH$), which are used as initiators for free-radical reactions. However, many substances can form peroxides on storage in contact with air, and such compounds may cause more problems than the peroxide reagents themselves because their presence may be unsuspected.

Organic peroxides are among the most hazardous substances handled in laboratories. Many of them are more sensitive to shock than most primary explosives such as trinitrotoluene (TNT). Although they are low-power explosives, they can detonate at very high rates, fragmenting glass vessels into small, high velocity particles. They are hazardous because many can be detonated by heat, friction, impact, or light, as well as by contact with strong reducing agents. Although each peroxide has a specific rate of decomposition under a given set of conditions, a low rate of decomposition can auto-accelerate because of poor heat dissipation and cause a violent explosion, especially in bulk quantities. Organic peroxides are highly flammable, and fires involving bulk quantities of them should be approached with extreme caution because of the possibility of an explosion.

Many ethers, acetals, olefins, and other common classes of organic compounds can form peroxides on storage in contact with air (Appendix B). Among common chemicals that can form dangerous concentrations of peroxides under these conditions are cyclohexene, p-dioxane, ethers, glyme, diglyme, tetrahydrofuran, and tetrahydronaphthalene. All solvents and reagents capable of producing peroxides should be dated at the time they are first opened for use (Figure 17) and should be discarded or tested for peroxide formation within a fixed time thereafter (Appendix B). Such compounds should be tested for peroxide content before they are used. The simplest qualitative test is based on oxidation of iodide ions to iodine. Add 1 cm^3 of the substance to be tested to a freshly prepared solution of 100 mg of sodium iodide or potassium iodide in 1 cm^3 of glacial acetic acid. A yellow colour indicates a low concentration of peroxide, and a brown colour indicates a high concentration. The test is sensitive to hydroperoxides (ROOH), which are the principal hazard of many solvents, but it does not detect those peroxides, such as dialkyl peroxides (ROOR), that are difficult to reduce. The latter may be detected using a reagent prepared by dissolving 3 g sodium iodide in 50 cm^3 of glacial acetic acid and adding 2 cm^3 of 37% hydrochloric acid. In addition, test paper containing a peroxidase is available that detects organic peroxides (including dialkyl peroxides) and oxidizing anions (e.g., sulfate(VII), (di)chromate(VI)) colorimetrically.

Hydroperoxides, the principal hazardous contaminants of peroxide-forming solvents, can be removed from such solvents by passage through

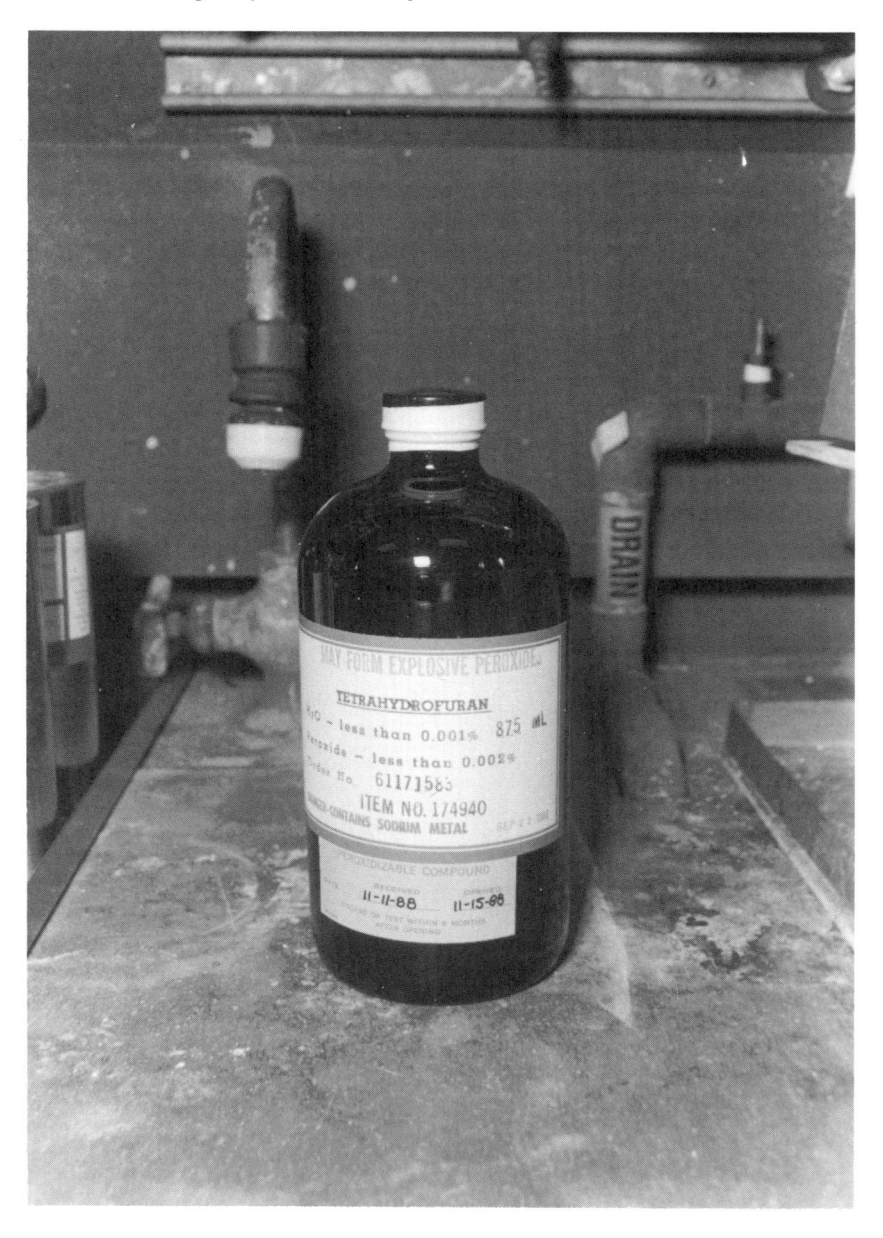

Fig. 17 Correct labelling of a bottle of peroxide-forming ether, with instruction to discard or test within 6 months after opening.

a column of basic activated alumina, by treating with an indicating molecular sieve or by reduction with iron(II) sulfate. Although these procedures do not remove dialkyl peroxides, they are usually present in such small quantities as not to be a significant hazard. A 2-cm x 33-cm column filled with 80 g of 80-mesh "F-20 Alcoa®" or "Woelm® basic activated alumina" is usually sufficient to treat 100-400 cm^3 of solvent, whether water-soluble or not. After passing through the column, the solvent should be retested to confirm that the peroxides have been removed. Peroxides formed by air oxidation are usually decomposed by the alumina, not merely adsorbed on it. However, as a safety measure, the wet alumina should be slurried with a dilute acidic solution of iron(II) sulfate before being discarded, or should be placed in a capped plastic jar or a polyethylene bag for incineration.

Hydroperoxides in 1 dm^3 of a water-insoluble solvent can be removed by treatment with a solution of 6 g of $FeSO_4.7H_2O$ and 6 cm^3 of concentrated sulfuric acid in 11 cm^3 of water. The solvent and solution are shaken together for a few minutes in a separating funnel or stirred vigorously in a flask until the solvent no longer gives a positive test for peroxide.

A new method for removing peroxides from ethers involves refluxing 100 cm^3 of the ether with 5 g of indicating, activated 4A molecular sieve pellets (4-8 mesh) for several hours under nitrogen. The sieve pellets are separated from the ether, and present no hazard because the hydroperoxides are destroyed during the operation, probably by catalytic action of the indicator in the pellets.

Many explosions have occurred during distillation of peroxide-containing substances, particularly when the distillation has been taken near to dryness. Laboratory operations with peroxides or peroxide-containing solvents should be carried out behind shields and by taking the following precautions.

1. The quantity of material should be limited to the minimum amount required. Unused peroxides should be destroyed and never returned to the container.
2. All spillages should be cleaned up immediately. Solutions of peroxides can be absorbed on vermiculite.
3. The sensitivity of most peroxides to shock and heat can be reduced by dilution with inert solvents, such as aliphatic hydrocarbons. Toluene is unsuitable for diacyl peroxides because it induces their decomposition.
4. Solutions of peroxides in volatile solvents should not be used under conditions in which significant quantities of the solvent might be vaporized because of the consequent increase in the peroxide concentration of the solution.

5. Metal spatulas must not be used to handle peroxides because contamination by metal can lead to explosive decomposition. Ceramic or inert plastic spatulas may be used.
6. Open flames and other sources of heat must not be used near peroxides.
7. Friction, grinding, and all forms of impact must be avoided in the presence of peroxides, especially solid ones. Old containers of peroxide-forming compounds are dangerous because solid peroxides may have formed in the threads of a screw-cap container and can detonate the entire contents of the container owing to friction when opening the cap. For this reason, peroxides should not be stored in glass containers that have screw-cap lids or glass stoppers. Polyethylene bottles that have plastic screw-cap lids can be used.
8. To minimize the rate of decomposition, peroxides should be stored at the lowest possible temperature consistent with their solubility or freezing point. However, liquid peroxides or solutions of peroxides should not be stored at or below the temperature at which the peroxide freezes or precipitates because frozen or precipitated peroxides are especially sensitive to shock.

The disposal of peroxides is discussed in Section 10.1.16.

7.2.2. Other explosive compounds

In general, compounds containing the following functional groups tend to be sensitive to heat and shock: acetylide, azide, haloamine, nitro, nitroso, ozonide, and diazo. For example, diazomethane may decompose explosively when a ground-glass joint in a vessel containing it is rotated. These and other examples of shock-sensitive compounds are listed in Appendix D.

Various combinations of common laboratory chemicals can produce explosions when they are brought together (e.g., acetone and chloroform in the presence of base). Some combinations can give explosive reaction products that explode long after mixing (e.g., ethanol and silver nitrate). Examples are listed in Appendix D.

7.2.3. Handling explosive compounds

Explosive chemicals decompose under conditions of mechanical shock, elevated temperature, or chemical action, releasing large volumes of gases, heat, toxic vapours, or combinations thereof. Government regulations cover the transportation, storage, and use of explosives. These regulations must be consulted before explosives and related dangerous materials are used in the laboratory.

Explosive materials should be brought into the laboratory only as required and then in the smallest quantities needed for the experiment. Explosives should be segregated from other materials that could create a serious danger to life or property should an accident occur.

The handling of highly energetic substances without injury demands attention to minute detail. The unusual nature of work involving such substances requires special safety measures and handling techniques that must be thoroughly understood and followed by all persons involved. The practices listed below are a guide for use in any laboratory operation that might involve explosive materials.

Personal protective apparel (Sections 2.1-2.3)

1. Safety glasses that have a cup-type side shield made of a light, clear plastic material affixed to the frame should be worn by all laboratory personnel.
2. A face shield that has a "snap-on" throat protector in place should be worn at all times when the worker is in an exposed position (e.g., when operating or manipulating synthesis systems, when bench shields are moved aside, or when handling or transporting explosive products).
3. Gloves should be worn whenever it is necessary to reach behind a shielded area while a hazardous experiment is in progress or when handling adducts or gaseous reactants. Yellow "electrical lineman's" gloves afford good protection against 2-g quantity detonations in glass, provided the detonation is at least 8 cm away. However, such a detonation in contact with a gloved hand would cause severe injury and possible loss of fingers.
4. Laboratory coats should be worn at all times. They should be of a slow-burning material and fitted with quick-release cloth buttons. These coats help to reduce minor injuries from flying glass and reduce the possibility of injury from an explosive flash.

Protective devices (Section 2.4)

Barriers such as shields, barricades, and guards should be used to protect personnel and equipment from injury or damage. The barrier should completely surround the danger area (Figure 18). An acrylic sliding shield 6 mm thick effectively protects a worker from glass fragments resulting from a laboratory-scale detonation. The shield should be closed when hazardous reactions are in progress or whenever hazardous materials are being temporarily stored. However, such shielding is not effective against metal shrapnel.

Working with flammabile or explosive substances 123

Fig. 18 Safety shield affording local protection in conducting a hazardous procedure.

Dry boxes should be fitted with safety-glass windows overlaid with acrylic sheet 6 mm thick. This protection is adequate against a laboratory-scale detonation. The problem of hand protection however still remains, although electrical lineman's gloves over the rubber dry box gloves offer some additional protection. Other safety devices should be used in conjunction with the gloves, for example, tongs and clamps.

Armoured hoods or barricades made with extra thick (2.5 cm) polyvinyl butyral resin shielding and heavy metal walls give complete protection against detonations of not over 20 g of an explosive. These hoods are designed for use with 100 g of material, but an arbitrary 20-g limit is usually set because of the noise level in the event of a detonation. Such hoods should be equipped with mechanical hands that enable remote manipulation of equipment and reagent containers (Figure 19). A sign should be posted such as: *CAUTION: NO ONE MAY ENTER AN ARMOURED HOOD FOR ANY REASON DURING THE COURSE OF A DANGEROUS OPERATION.*

Note that these armoured hoods are designed for manipulations of potentially explosive chemicals; they should not be used as barricades for protection against the projectiles and massive release of gas that can result from failure of pressure equipment (Section 8.2.3).

Other protective devices such as long- or short-handled tongs for holding or manipulating dangerous items at a safe distance, and remote control equipment such as mechanical arms, stopcock turners, lab jack turners, and remote cable controllers, should be available as required to prevent exposure of any part of the body to injury.

Reaction quantities

In conventional explosives laboratories, no more than 0.5 g of product should be prepared in a single reaction. During the reaction period, no more than 2 g of reactants should be present in the reaction vessel. Thus the diluent, the substrate, and the energetic reactant must all be considered when determining the total explosive power of the reaction mixture. Special reviews should be established to examine operational and safety problems involved in scaling-up a reaction in which an explosive substance is used or formed.

Reaction operations

Various heating methods may be used, the most common being heating tapes or mantles, and sand, water, steam, or silicone oil baths. Heat guns can be used for certain operations, but their use must be prohibited when a flammable vapour is present. All controls for heating and stirring equipment should be operable from outside the shielded area.

Vacuum pumps (Section 4.2) should carry tags indicating the date of the most recent oil change. Oil should be changed once a month, or sooner if

Fig. 19 Remote control of a high-pressure reaction vessel, with a steel baricade facility.

it is known that it has been exposed to reactive gases. All pumps should either be vented into a hood or trapped. Vent lines may be of Tygon®, rubber, or copper. If Tygon®, or rubber lines are used, they should be supported so that they do not sag.

When potentially explosive materials are being handled, only very small quantities should be used. A shielded Dewar flask (Section 8.3.2) should be used for condensing volatile explosive materials. Heating baths of flammable materials should not be used.

7.3. OTHER CHEMICAL HAZARDS THAT CAN LEAD TO FIRES OR EXPLOSIONS

7.3.1. Hazardous substances or materials

Some of the most common laboratory sources of fires and explosions are listed here:

1. *Acetylenic compounds.* Acetylene (Section 8.4.1) is explosive in mixtures of 2.5-80% (v/v) with air. At pressures above 2 atmospheres it decomposes with explosive violence if subjected to an electrical discharge or high temperature. Some substituted acetylenes are equally dangerous. Dry metal acetylides can detonate on receiving even a slight shock.

2. *Aluminium chloride* is dangerous because of its violent reaction with water or other compounds containing active hydrogen. Moisture in a closed container may cause sufficient hydrogen chloride to be evolved so as to build-up considerable pressure. If a bottle is to be opened after long standing, it should be completely enclosed in a heavy dry towel.

3. *Ammonia* (Section 8.4.2) reacts with iodine to give nitrogen triiodide, and with chlorates(I) to give hydrazine or nitrogen chlorides, all of which are explosive and highly toxic. Mixtures of ammonia and organic halides sometimes react violently when heated under pressure.

4. *Dry benzoyl peroxide* (Section 7.2.1) is easily ignited and sensitive to shock. It decomposes spontaneously at temperatures above 50 $^{\circ}$C. It is reportedly desensitized by addition of 20% water.

5. *Carbon disulfide* is very toxic and flammable. Its vapours in air can be ignited by a steam bath, a hot-plate, a hot water pipe or a glowing light bulb.

6. *Chlorine* (Section 8.4.5) may react violently with hydrogen or hydrocarbons when exposed to sunlight. Chlorine should never be used in aluminium equipment.

7. *Chromium(VI) oxide-pyridine complex* ($CrO_3 \cdot C_5H_5N$) may explode if the CrO_3 concentration is too high. The complex should be prepared by cautious addition of CrO_3 to excess pyridine.
8. *Diazomethane* and related compounds should be treated with extreme caution. They are very toxic, and the pure gases and liquids explode readily.
9. *Dimethyl sulfoxide* decomposes violently on contact with a wide variety of active halogen compounds or metal hydrides. It penetrates the skin readily, and can carry solutes or chemicals on the skin into the body. It is moderately toxic.
10. *"Dry ice"* (solid carbon dioxide) should not be kept in a closed container unless it is designed to withstand pressure. Poorly sealed containers of other substances stored over dry ice for extended periods absorb carbon dioxide, and if they are removed from storage and allowed to come to room temperature rapidly, the carbon dioxide may develop sufficient pressure to rupture the container. When such containers are removed from storage, the stopper should be loosened or the container wrapped in towels and kept behind a shield. Dry ice and other cryogenic substances can produce serious cold burns.
11. *Drying agents* such as sodium hydroxide on an inert support should not be mixed with phosphorus(V) oxide because the mixture may explode if it is warmed with a trace of water. Drying agents that contain cobalt salts as moisture indicators should be used only with gases because cobalt salts can be extracted by some organic solvents.
12. *Ethylene oxide* has been known to explode when heated in a closed vessel, and all experiments under pressure should be carried out behind suitable barricades (Section 7.2.3).
13. *Halogenated compounds*, such as chloroform, carbon tetrachloride and methylene chloride, should never be dried with sodium, potassium, or other active metals because the mixture is potentially explosive. Many halogenated compounds are toxic.
14. *Hydrogen peroxide* in aqueous solutions can be dangerous; such a solution can cause severe skin burns. Aqueous 30% H_2O_2 may decompose violently if contaminated with iron, copper, chromium, or other heavy metals or their salts.
15. *Liquid-nitrogen* cooled traps can condense oxygen from the air into an open container (Section 7.1.3). If the coolant is removed and the container closed, an explosion can result

from pressure build-up in the container, or from reaction of its contents with the oxygen. Accordingly, only sealed or evacuated equipment should be so cooled (Section 8.3.4).

16. *Lithium aluminium hydride* (LiAlH$_4$) should not be used to dry methyl ethers or tetrahydrofuran because of a fire hazard. The products of its reaction with CO_2 are reported to be explosive. Accordingly, carbon dioxide or hydrogen carbonate extinguishers should not be used on LiAlH$_4$ fires; such fires should be smothered with sand or some other inert substance.

17. *Manganate(VII)* in contact with concentrated sulfuric acid can form the highly explosive oxide Mn_2O_7.

18. *Oxygen* (Section 8.4.10). Serious explosions have resulted from contact of mineral-based oils with high pressure oxygen. Oil should never be used on the connections to an oxygen cylinder.

19. *Ozone* is a toxic, highly reactive gas. It is formed by the action of ultraviolet light on oxygen, so that air exposed to an ultraviolet source should be vented to the exhaust hood.

20. *Palladium, platinum, platinum oxide, Raney nickel, and other hydrogenation catalysts* should be filtered carefully from hydrogenation reaction mixtures because the filtered catalyst is saturated with hydrogen and highly reactive. It may inflame spontaneously on exposure to air. The filter cake should not be allowed to become dry, particularly on a large-scale. The funnel containing the moist filter cake should be put into water immediately after the filtration. Such catalysts should not be added to a flask in the presence of hydrogen because of the danger of an explosion.

21. *Chlorates(VII)* are coming into increasing use in many areas of chemistry. However, "heavy metal" and all organic salts of chloric(VII) acid are highly explosive, and should always be handled as explosive materials. They are quite sensitive to heat, shock, and friction. All manipulations of chloric(VII) acid or chlorates(VII) should be carried out with the precautions prescribed for explosive compounds. Chlorates(VII) are potent oxidizing agents, and should never be used as drying agents for organic compounds or with a dehydrating acid strong enough to concentrate the chloric(VII) acid beyond 70%, for example, in a drying flask that has a sulfuric acid bubbler.

22. *Aqueous 70% chloric(VII) acid* can be boiled safely at approximately 200 °C, but neither the acid nor its vapour should be allowed to come into contact with any organic substance or

Working with flammabile or explosive substances 129

with any inorganic reducing agent such as Sb(III). Evaporation of chloric(VII) acid should only be carried out in a hood with a good draught and no wood construction in the flue. If this acid is being used regularly, the hood and ducts should be washed weekly to avoid the danger of spontaneous combustion or explosion. Disposal of chlorates(VII) is discussed in Chapter 11.

23. *Peroxides (inorganic)*. Peroxides of barium, sodium, or potassium form explosive combinations with organic materials.
24. *Phosphorus (red or white)* forms an explosive mixture with oxidizing agents. White phosphorus should be stored under water because it ignites spontaneously in air. The reaction of phosphorus with aqueous hydroxides produces phosphine, which is toxic and ignites spontaneously or explodes in air.
25. *Phosphorus(III)* chloride reacts with water to form phosphoric(III) acid. Care should be taken in opening containers of phosphorus(III) chloride and samples that have been exposed to moisture should not be heated without adequate shielding.
26. *Potassium metal* is more reactive than sodium. It ignites quickly on exposure to humid air and should therefore be handled under the surface of a hydrocarbon solvent such as mineral oil or toluene (see entry for sodium below). Old samples of potassium metal may become covered with a yellow crust of peroxides, which are shock-sensitive. Such samples should never be cut, even under the surface of a hydrocarbon (Section 10.2.4).
27. *Sodium metal* should be stored in a closed container under kerosene, toluene, or mineral oil. Contact with water must be avoided because the metal reacts violently to form hydrogen with evolution of sufficient heat to cause ignition. Carbon dioxide, sodium hydrogen carbonate, or halocarbon fire extinguishers should not be used on alkali metal fires.
28. *Sulfuric acid* should be avoided if possible as a drying agent in desiccators. However, if it is used, glass beads should be placed in the desiccator to avoid splashing when the desiccator is moved. Sulfuric acid must not be used in melting point baths; silicone oil is just as effective and much safer. When diluting sulfuric acid, always add the acid slowly to cold water, never the reverse.
29. *Trichloroethylene* reacts with sodium hydroxide or potassium hydroxide to form dichloroacetylene, which ignites spontaneously in air and detonates readily even at 70 °C. Trichlo-

roethylene is toxic, and suitable precautions should be taken when it is used as a degreasing solvent.
30. *Vacuum-distillation residues* have been known to explode when the still is vented to the air before the residue is cooled. Such explosions can be avoided by venting the still pot with nitrogen, or by cooling it to room temperature before venting with air.

7.3.2. Incompatible chemicals

Care must be taken when handling or disposing of a chemical that it does not come in contact with another chemical with which it may react violently. Appendix E provides guidelines on classes of incompatible chemicals.

8

Procedures for Working with Gases at Pressures Above or Below Atmospheric

All cylinders of compressed gas are potentially dangerous because of the energy of compression built up in the cylinder. Many such gases are flammable, and have a low flash point. If any of them is allowed to diffuse through the laboratory, there is an imminent danger of fire or explosion. Many gases are chemically reactive or toxic. Even "harmless" gases such as nitrogen or carbon dioxide can cause asphyxiation by oxygen deprivation. As a result, detailed procedures are necessary for the safe handling of compressed gases, the cylinders that contain them, the regulators used to control their flow, and the piping used to transport them.

8.1. GAS CYLINDERS

Improperly constructed gas cylinders are potentially very dangerous; consequently many countries specify the materials to be used in cylinder construction and the capacities, test procedures, and service pressures to be used. Generally a compressed gas steel cylinder must have a designation showing under what government specification it was manufactured and the acceptable operating pressure (e.g., 14 MPa, at 21 $^{\circ}$C).

The physical state of the material within a cylinder determines the pressure in it. In the absence of a liquid phase, the pressure is determined by the quantity of gas compressed into the cylinder volume, and will decrease as gas is withdrawn from the cylinder. On the other hand, liquefied gases such as propane or ammonia will exert their own vapour pressure as long as any liquid remains in the cylinder, if the critical temperature of the compound is not exceeded.

Prudent practices for the use of compressed gas cylinders in the laboratory include proper identification, transportation, storage (Figure 20), handling and use (Sections 5.2.6, 5.3.2), and return of the empty cylinder.

8.1.1. Identification

The contents of any compressed gas cylinder should be clearly identified so as to be easily, quickly, and completely determined by any laboratory worker. Such identification should be stencilled or stamped on the cylinder itself, or a label should be provided that cannot be removed from the cylinder. Three-part tag systems, which are available commercially, can be useful for identification and inventory. No compressed gas cylinder should be accepted that does not identify its contents legibly by name (Section 5.1). Colour coding is not a reliable means of identification. Cylinder colours vary from one supplier to another, and labels on caps have no value because caps may be interchangeable. If the labelling on a cylinder becomes unclear or an attached tag is so defaced that the contents cannot be identified, the cylinder should be marked "contents unknown" and returned directly to the supplier.

All gas lines leading from a central compressed gas supply should be clearly labelled to identify the gas and the laboratory served, and be marked with emergency telephone numbers in case of accidents. The labels should be colour coded to distinguish flammable, toxic, or corrosive gases (e.g., a yellow background and black letters) from inert gases (e.g., a green background and black letters).

Signs should be conspicuously posted in areas in which flammable compressed gases are stored, identifying the substances and appropriate precautions, e.g., *HYDROGEN - NO SMOKING - NO OPEN FLAMES*.

8.1.2. Handling and use of cylinders

Compressed gas cylinders should be firmly secured at all times using a clamp and belt or chain to prevent their being knocked over with the possible risk of breaking the valve and releasing gas. Large cylinders should always be moved, strapped, or cradled on a two-wheel truck, never angled and rolled on the base (Figure 21).

Standard cylinder-valve outlet connections should be used to prevent the mixing of incompatible gases due to an interchange of connections. In general, right-handed threads are used for non-fuel and water-pumped gases, and left-handed threads for fuel and oil-pumped gases. Information on the standard equipment assemblies for specific compressed gases is available from the supplier. The assembly of miscellaneous parts, even of recognized approved types, could create a hazard. The threads on cylinder valves, regulators, and other fittings should be examined to ensure that they correspond to one another and are undamaged.

A gas cylinder under a pressure of more than 690 kPa should never be used without a pressure-reducing valve. Cylinders should be placed so that the valve and the pressure-reducing valve are readily accessible. The main cylinder valve should always be kept closed except when gas is being

Working above or below atmospheric pressures 133

Fig. 20 Safe storage of gas cylinders outside a laboratory building, with clearly labelled pipeline manifolds.

Fig. 21 Gas cylinder, on stand, securely attached to bench top.

withdrawn. This precaution is necessary not only for safety in the event that the pressure-reducing valve fails, but also to prevent the corrosion and contamination that could result from diffusion of air and moisture into the cylinder after it has been emptied.

Most cylinders are equipped with hand wheel valves. Those that are not should have a spindle key on the valve spindle or stem while the cylinder is in service. Only a wrench (spanner) or other tool provided by the cylinder supplier should be used to open a valve. Pliers or a pipe wrench should never be used. A valve may require a washer; this should be checked before the regulator is fitted.

Cylinder valves should be opened slowly to avoid the possibility of breaking the pressure-regulating valve by gas impact. It is rarely necessary or desirable to open the main cylinder valve all the way; the resulting flow would be much greater than needed. When opening the valve on any cylinder, the user should stand on the side of the cylinder opposite the valve port.

Disposal of leaking or unidentified cylinders is outlined in Section 10.5.

8.1.3. Pressure regulators

A pressure regulator is used to reduce a high pressure gas to a lower delivery pressure and flow rate for the required operating conditions. Regulators can be obtained with a range of supply and delivery pressures, flow capacities, and construction materials. Under no circumstances should oil be used on regulator valves or cylinder valves. Each regulator should be supplied with a specific standard inlet connection to fit the outlet connection on the cylinder valve for the particular gas.

Different types of regulators are used with different gases, and only the type of regulator specified by the vendor should be used. Regulators for non-corrosive gases are usually made of brass. Regulators made of corrosion-resistant materials are used for such gases as ammonia, boron(III) fluoride, chlorine, hydrogen chloride, hydrogen sulfide, and sulfur(IV) oxide. Vendors sell heaters as an accessory to avoid liquefaction inside the regulators for carbon dioxide and other easily liquefied gases.

All pressure regulators should be equipped with spring-loaded pressure relief valves to protect the low pressure side. If a cylinder of a dangerous gas is fitted with a regulator that has a built-in relief valve and is used in the laboratory, the regulator should have an external fitment for the attachment of a pipe to vent any escaping gas into the back of a hood. Some laboratories deal with the venting problem by placing the cylinders in a well vented shed outside the laboratory, and piping the gas into the laboratory. A two-stage regulator may be equipped with an internal bleed device to protect the low pressure stage from full cylinder pressure in case the high-pressure stage fails. When such regulators are used on cylinders containing dangerous gases, they should be fitted with a pipe for venting the gas into a hood or other safe location.

8.1.4. Flammable gases

Sparks and flames should be kept from the vicinity of cylinders of flammable gases. An open flame should never be used to search for leaks of flammable gases. Soapy water should be used, except during freezing weather, when a 50% glycerine-water solution or its equivalent may be used. Connections to piping, regulators, and other appliances should always be kept tight to prevent leakage, and the hoses kept in good condition. If a flammable gas is being used from a cylinder or other pressure equipment at a high flow rate, the cylinder must be connected to electrical ground with a cable to reduce the hazard of static spark ignition.

Regulators, hoses, and other appliances used with cylinders of flammable gases must not be interchanged with similar equipment intended for use with other gases.

All cylinders should be stored in a well ventilated place. Cylinders of flammable gases should never be stored in the vicinity of cylinders containing oxygen (Section 5.2.6).

8.1.5. Empty Cylinders

A cylinder should never be emptied to a pressure lower than 170 kPa because it may become contaminated if it is emptied completely and the valve left open. "Empty" vendor-owned cylinders should never be refilled at the laboratory; instead the regulator should be removed and the valve cap replaced. The cylinder should be clearly marked as "empty" and returned to a storage area for pickup by the supplier. Empty and full cylinders should not be stored in the same place.

Cylinder discharge lines in a central gas supply system should be equipped with approved check valves to prevent inadvertent contamination of other cylinders if there is a possibility of flow reversal. Sucking back is especially likely when gases are being used as reactants in a closed-system. A cylinder in such a system should be shut-off and removed when the pressure falls to 170 kPa. If there is the possibility that a cylinder has been contaminated, it should be so labelled and returned to the supplier.

8.2. NON-VENDOR-OWNED PRESSURE VESSELS AND OTHER EQUIPMENT

8.2.1. Records, inspection, and testing

Small laboratory-owned cylinders, which are often used for special gases, should be refilled only by trained and qualified personnel. Each pressure vessel should have its allowable working pressure and temperature (at this pressure), and the material of construction stamped on it or on an attached plate. Similarly, the relieving pressure and setting data should be stamped on a metal tag to be installed on a pressure relief device. The setting mechanism should be sealed. Relief devices used on pressure regulators do not require these seals or numbers.

All pressure equipment should be tested and inspected regularly. The frequency depends upon the extent of usage and the gases involved. Corrosive or other similar dangerous gases require more frequent tests and inspections. The identity of the inspector and the date of the latest inspection should be stamped on, or attached to, the equipment.

The assembled apparatus should be tested for leaks by pressurizing it with air or nitrogen to the maximum allowable working pressure and applying soap solution to threaded joints, packings, valves, and other connections. Before any pressure equipment is altered, repaired, stored, or shipped it should be vented carefully, and all toxic or other hazardous material removed completely so that it can be handled safely. Especially hazardous materials may require special cleaning techniques.

8.2.2. Assembly and operation

During the assembly of pressure equipment and piping, care should be taken to avoid strains and concealed fractures resulting from the use of improper tools or excessive force. Piping should not be used to support equipment of any significant weight.

Threads that do not fit accurately must not be forced. Thread connections should match correctly; tapered pipe threads cannot be joined with parallel machine threads. Parts with damaged or partly stripped threads should be rejected. Motor oil is a suitable thread lubricant for equipment to be used with non-oxidizing gases, but not with oxidizing gases such as oxygen or fluorine. Molybdenum(IV) sulfide powder or a fluorinated lubricant can be used as a thread lubricant with oxidizing gases.

In assembling copper tubing installations, sharp bends should be avoided and considerable flexibility should be allowed. Copper tubing hardens and cracks on repeated bending or flexing. It should be inspected frequently for leaks and fractures and renewed when necessary.

Stuffing boxes and gland joints are a source of problems in pressure installations. Particular attention should be given to the proper installation and maintenance of these parts, including the correct choice of lubricant and packing material. Light petroleum grease is a satisfactory lubricant for glands and stuffing boxes used with non-oxidizing gases. However, Teflon® disks, rings, or other purpose designed packing materials must be used when oxygen or strongly oxidizing gases are being handled.

Experiments in closed-systems that involve highly reactive materials, such as those subject to rapid polymerization (e.g., butadiene, methyl acrylate, acrylonitrile), should be preceded by small-scale tests using the same reaction materials to assess the possibility of an unexpectedly rapid reaction or unforeseen side reaction.

Autoclaves and other pressure-reaction vessels should not be filled more than half-full to ensure that space remains for expansion of the liquid when it is heated. Leak corrections or adjustments to the apparatus should not be made while it is pressurized. The system should be depressurized before mechanical adjustments are made.

Where low pressure equipment has been connected to a high-pressure source, the low pressure equipment should be disconnected entirely, or left independently vented to the atmosphere at the end of the experiment. This will prevent the gradual build up of excessive pressure in the low pressure equipment due to leakage from the high-pressure side.

Vessels or equipment made partly or entirely of silver, copper, or alloys containing more than 50% copper should not be used in contact with acetylene or ammonia. Equipment that contains metals susceptible to amalgamation (e.g., copper, brass, zinc, tin, silver, lead, gold) must not come into contact with mercury. This precaution includes equipment that has soldered and brazed joints.

Prominent warning signs should be placed in any area where a pressure reaction is in progress so that others entering the area will be aware of the potential hazard.

8.2.3. Barricades

All reactions under pressure must be isolated by barriers. Bench shields (Section 2.4.4) suffice for small scale hydrogenations and other reactions up to a pressure of about 300 kPa. Barricades with remote controls are needed at higher pressures. Moreover, any reaction on a scale greater than 1 g where there is the possibility of detonation should be carried out behind a properly designed barricade.

Steel barricades should be designed to serve three purposes: to contain missiles in the event of equipment failure; to reduce the sound levels generated by a decomposition; and to provide effective ventilation for dilution of toxic or explosive vapours that may be released. A TNT-equivalent rating can be calculated for a barricade and used to estimate the permissible scale of an experiment at which the barricade will contain missiles and suppress sound to a permissible level. Critical features of barricade design comprise:

1. A properly designed vent or a weak section that opens to a relief area which will facilitate the release of pressure without allowing missiles to escape and cause injury to personnel. *IT IS AN ESSENTIAL FEATURE OF ANY BARRICADE*, and serious accidents have occurred for the lack of it.

2. Lightweight covering of the relief area to maintain exhaust ventilating patterns and capacity.
3. Missile-resistant windows.
4. Outside access to key controls such as valves and switches.
5. Spark-free, heavy duty electrical service outlets.
6. Equipment for fire suppression.
7. Alarms for reaction conditions such as unduly high or low temperature and pressure.
8. Safety interlocks to shutdown the reaction in case of ventilation failure.
9. Fail-safe designs for certain valves to fail in the appropriate position (open or closed) if electrical power or air control should be lost.
10. Provision for automatic blanketing or purging with inert gas in the event of equipment failure. Any such automatic inert gas system should be equipped with an audible alarm to warn people of a possibly oxygen-deficient atmosphere.
11. Catch drums to retain liquid discharged from relief devices.
12. Scrubbers in exhaust ducts.

8.2.4. Pressure relief devices

All pressure or vacuum systems and all vessels that may be subjected to pressure or vacuum should be protected by pressure relief devices. However, it should be borne in mind that a pressure relief device may not respond quickly enough to prevent fragmentation of a reactor in the event of a detonation. Consequently, any reaction that might detonate should be carried out behind a barricade. Examples of pressure relief devices include the rupture disks used with closed-system vessels and the spring-loaded relief valves used with vessels for transferring liquefied gases. The following considerations are advisable in the use of pressure relief devices:

1. The maximum setting of a pressure relief device is the rated working pressure established for the vessel, or for the weakest component of the pressure system, at the operating temperature. The operating pressure should be less than the allowable working pressure of the system. The maximum operating pressure of a system protected by a spring-loaded relief device should be from 5 to 25% lower than the rated working pressure, depending on the type of safety valve and the importance of leak-free operation. The maximum operating pressure of a system protected by a rupture disk device should be about two-thirds of the rated working pressure. The exact figure is gov-

erned by the fatigue life of the disk, the temperature, and load pulsations.
2. Pressure relief devices that may discharge toxic, corrosive, flammable, or otherwise hazardous or noxious materials should be vented to a safe disposal area. It is essential that pipe connections between a rupture disk and a pressure vessel, as well as piping to vent a rupture disk, be straight and not reduced in size. Even $45°$ curves in such piping can impede the flow of gas if the rupture disk blows, resulting in back pressure and a possible secondary explosion.
3. Shut-off valves should not be installed between pressure relief devices and the equipment they are to protect.
4. Only qualified persons should perform maintenance work on pressure relief devices.
5. Pressure relief devices should be inspected periodically.
6. A spring-loaded pressure relief device should never be tied down, for example because it is leaking.

8.2.5. Pressure gauges

The proper choice and use of a pressure gauge is the responsibility of the user. Among the factors to be considered are the flammability, compressibility, corrosiveness, toxicity, temperature, and pressure range of the fluid with which it is to be used.

A pressure gauge is necessarily a weak point in a pressure system because its measuring element must operate in the elastic zone of the metal involved. The resulting limited factor of safety makes the correct gauge selection important and often dictates the use of other accessory protective equipment. The most commonly used gauge is a Bourdon tube, which is usually made of brass or bronze and has soft-soldered connections. More expensive gauges can have Bourdon tubes made of steel, stainless steel, or other special metals with welded or silver-soldered connections. Accuracies vary from 2% for less-expensive pressure gauges to 0.1% for higher quality gauges. These tolerances apply to the middle half or three-quarters of the gauge range.

Alternative pressure measuring devices, such as the strain-gauge transducer, may provide greater safety and these are being used more commonly.

8.2.6. Glass equipment

The use of glassware for work at superatmospheric pressure should be avoided whenever possible. Glass is a brittle material subject to unexpected failure from mechanical impact or assembly stresses. Glass equipment that is incorporated in metal pressure systems, such as rotameters and liquid level gauges, should be installed with shut-off valves at both ends to control the discharge of liquid or gaseous materials in the event of breakage.

Glass equipment in pressure or vacuum service should be provided with adequate shielding to protect users from flying glass and the contents of the equipment. New or repaired glass equipment for pressure or vacuum work should be examined for flaws and strains under polarized light.

Corks, rubber stoppers, and rubber or plastic tubing should not be relied on as relief devices for glass equipment. A liquid seal, Bunsen tube, or equivalent positive relief device should be used.

8.2.7. Plastic equipment

Except as noted below, the use of plastic equipment for pressure or vacuum work should be avoided unless no substitute is available.

Tygon® and similar plastic tubings have some limited applications in pressure work. These materials can be used with natural gas, hydrocarbons, and most aqueous solutions at room temperature and moderate pressure. Details of permissible operating conditions must be obtained from the manufacturer. Because of their very large coefficients of thermal expansion, some polymers have a tendency to expand a great deal on heating. Thus if a valve or joint is tightened when the apparatus is cold, the plastic can occlude when the temperature increases. This can be a hazard where equipment is subjected to very low temperatures or to alternating low and high temperatures.

8.2.8. Piping, tubing, and fittings

All-brass fittings should be used with copper or brass tubing, and steel or stainless steel fittings with steel or stainless steel tubing. It is important that fittings of this type be installed correctly. It is inadvisable to mix different types of fittings in the same apparatus assembly, e.g., brass with steel.

8.2.9. Sealed tube reactions

A high-pressure autoclave should be used for any reaction on a scale greater than 10-20 g of reactants or at a pressure above 690 kPa. However, it is sometimes convenient to run small-scale reactions at low pressures in a small sealed glass tube or in a thick-walled pressure bottle of the type used for catalytic hydrogenation. Sealed tube experiments are inherently more dangerous than properly conducted experiments in autoclaves, and should be avoided if possible. The laboratory worker should be fully prepared for the possibility that a sealed vessel may burst. Every precaution should be taken to avoid injury from flying glass or from hot, corrosive, or toxic reactants. Centrifuge bottles used for pressure reactions should be sealed with rubber stoppers clamped in place, wrapped with friction tape, surrounded by multiple layers of loose cloth towelling, and clamped behind a good safety shield. The preferred source of heat for such vessels is steam, because an explosion in the vicinity of an electrical heater could start a fire, and an explosion in a liquid heating bath would distribute hot liquid. Any reaction of this type should be labelled with signs that indicate the contents of the reaction vessel and the explosion hazard.

Similar precautions should be followed for reactions in sealed tubes. The sealed glass tubes can be placed inside pieces of brass or iron pipe capped at one end with a pipe cap, or alternatively in an autoclave containing some of the reaction solvent to equalize the pressure inside and outside the glass tube. The tubes can be heated with steam or in a specially constructed, electrically-heated "sealed-tube" furnace that is thermostatically controlled and located so that the force of any explosion is directed into a safe area.

Small tubes can be made of ordinary Pyrex® glass with an outside diameter of 5-8 mm. Special thick-walled Pyrex® pressure tubing must be used for larger diameter reaction vessels. The reaction tube must be constructed with care to avoid thin spots in the bottom or in the constricted section. The tube should not be more than half-full. The reactants should be introduced with a pipette or syringe in such a way that no material is deposited on the constricted section of the tube. The tube is then purged with nitrogen while being cooled in crushed dry ice or a mixture of dry ice with a non-flammable liquid such as methylene chloride (hood); a dry ice-acetone bath should not be used because of the fire hazard. When the reactants are thoroughly chilled, the nitrogen supply is removed and the constricted section is immediately sealed using a small oxygen gas torch. The sealed tube should be heated by steam or in a "sealed-tube" furnace.

When the required heating has been completed, the sealed tube or bottle should be allowed to cool to room temperature. If a sealed-tube furnace has been used and the contents of the tube are non-flammable, the tube may be opened by heating the tip protruding from the furnace with a small oxygen-gas flame until the pressure is released by breakage of the molten glass. Sealed bottles and tubes of flammable materials should be wrapped with cloth towelling, placed behind a safety shield, and then slowly cooled, first in an ice bath and then in dry ice, after which the clamps and rubber stoppers can be removed from the bottles and the tips of tubes heated to the melting point to release any remaining internal pressure. After the pressure has been released, the tubes can be cut open by cracking along a file mark. Alternatively, pressure may be released by marking the top of the tube with a sharp file and cracking the tip. This operation must be carried out behind a transparent shield while wearing heavy gloves, a face shield and apron.

8.2.10. Handling of liquefied gases and cryogenic liquids

The primary hazards of cryogenic liquids are fire or explosion, pressure build-up, embrittlement of structural materials, contact with and destruction of living tissue, and asphyxiation.

The fire or explosion hazard is obvious when gases such as hydrogen, methane, or acetylene are used. Enriched oxygen will greatly increase the flammability of ordinary combustible materials and may even cause some normally non-combustible materials such as carbon steel to burn.

Oxygen-saturated wood and asphalt have been known to explode when subjected to shock. Because oxygen has a higher boiling point (-183 $^{\circ}$C) than nitrogen (-191 $^{\circ}$C), hydrogen (-253 $^{\circ}$C), or helium (-269 $^{\circ}$C), it can be condensed out of the atmosphere during the use of these lower-boiling cryogenic liquids. Explosive conditions may be created, particularly with the use of liquid hydrogen (Section 8.4.7).

Even brief skin contact with a cryogenic liquid is capable of causing tissue damage similar to that of thermal burns, and prolonged contact may result in serious tissue damage.

Eye protection, preferably a face shield, should be worn when handling liquefied gases and other cryogenic fluids (Section 2.1). Gloves should be chosen that are inert and impervious to the fluid being handled and loose enough to be easily removed (Section 2.2). The working area should be well ventilated. The transfer of liquefied gases from one container to another should not be attempted for the first time without direct supervision and instruction from an experienced worker.

Cylinders and other pressure vessels used for the storage and handling of liquefied gases should not be filled to near the vessel's water capacity. Each liquefiable gas has its own safe filling density, i.e., the maximum weight of material permitted in a cylinder as a percentage of the weight of water the cylinder will hold. The allowable filling density depends on the gas and on the pressure rating of the cylinder. Compressed gas vendors and regulatory agencies publish lists of safe filling densities for loading various gases into cylinders of given pressure ratings.

8.2.11. Low temperature equipment

The impact strength of ordinary carbon steel is greatly reduced at low temperatures. The steel may fail when subjected to impact or mechanical shock, even though its ability to withstand slowly applied loading is not impaired. This type of failure normally occurs at points of high stress, such as at notches in the material or abrupt changes of cross-section.

The 18% chromium-8% nickel stainless steels (type 304) retain their impact resistance down to approximately -240 $^{\circ}$C, the exact value depending upon design. The impact resistance of aluminium, copper, nickel, and many other non-ferrous metals and alloys increases with decreasing temperatures.

8.2.12. Hydrogen embrittlement

Special alloy steels must be used in working with liquids or gases containing hydrogen at temperatures greater than 200 °C or at pressures greater than 21 MPa because of the risk of weakening carbon steel equipment by hydrogen embrittlement.

8.3. VACUUM WORK

In an evacuated system, the higher pressure is on the outside rather than on the inside, so that a break causes an implosion rather than an explosion. The resulting hazards include flying glass, chemicals, and possibly fire.

A moderate vacuum, such as 10 mm Hg, which can be achieved by a water aspirator, may seem safe compared with a high vacuum such as 10^{-5} mm Hg. However, the pressure differences between outside and inside are comparable (760 - 10 = 750 mm Hg in the first instance, compared with 760 - 10^{-5} = 760 mm Hg in the second instance). Therefore, any evacuated container must be regarded as an implosion hazard.

Vacuum-distillation apparatus often provides some of its own protection in the form of heating mantles and column insulation. However, this is not sufficient protection, because an implosion would scatter hot, flammable liquid. An explosion shield should be used to protect the worker (Section 2.1).

Equipment at reduced pressure is especially prone to rapid changes in pressure, within the apparatus. This may result in liquid reactants moving from one part of the apparatus to another, with undesirable and possibly dangerous consequences.

Water, solvents, or corrosive gases must not be drawn into a vacuum system. When the potential for such a problem exists, a water-aspirator should be used as the vacuum source.

A mechanical vacuum pump should be protected by placing cold traps (Section 8.3.4) between the pump and the equipment, and the exhaust should be vented to an exhaust hood or to the outside of the building (Section 4.2). If solvents or corrosive substances are inadvertently drawn into the pump, the oil should be changed promptly before any further use. The belts and pulleys on vacuum pumps should be covered with guards.

8.3.1. Glass vessels

Glass vessels at reduced pressure can collapse violently from an accidental blow or if already cracked or weakened. Therefore vacuum operations must be conducted behind adequate shielding. It is advisable to check for flaws such as star cracks, scratches, or etching marks each time glass vacuum apparatus is used. Only round-bottomed or thick-walled, flat-bottomed flasks (e.g., Pyrex®) specifically designed for operations at reduced pressure should be used as reaction vessels. Repaired glassware should be checked for strains by viewing with crossed polarizing filters before being used for operations at reduced pressure.

8.3.2. Dewar flasks

These flasks are capable of collapsing as a result of thermal shock or of a very slight scratch by a stirring rod. They should be shielded by a layer of friction tape or by enclosure in a wooden or metal container to reduce the hazard of flying glass in case of collapse.

8.3.3. Desiccators

Glass vacuum desiccators should be made of Pyrex® or similar glass. They should be completely enclosed in a shield or wrapped with friction tape in a grid pattern leaving the contents visible, but at the same time guarding against flying glass. Various plastic (e.g., polycarbonate) desiccators now on the market reduce the implosion hazard and may be preferable.

8.3.4. Cold traps

Cold traps interposed between a vacuum system and a vacuum pump should be of sufficient size and at low enough temperature to collect all condensable vapours. They should be checked frequently to ensure they do not become blocked by the frozen material collected in them.

The common practice of using acetone and "dry ice" as a coolant should be avoided. Isopropyl alcohol or ethanol work as well as acetone and are less flammable, and less prone to foam on addition of small particles of "dry ice". "Dry ice" and liquefied gases used in refrigerant baths should always be open to the atmosphere, as a closed-system could develop uncontrolled and dangerously high pressure.

After completion of an operation in which a cold trap has been used, the system should be vented. This is important because volatile substances that have collected in the trap may vaporize when the coolant has evaporated and cause a pressure build-up. In addition, oil from the vacuum pump may be sucked back into the system.

Extreme caution should be exercised in using liquid nitrogen as a coolant for a cold trap. If such a system is opened while the cooling bath is still in contact with the trap, oxygen may condense from the atmosphere, and if the trap contains organic material, a highly explosive mixture may form. Thus a system that is connected to a liquid-nitrogen trap should not be opened to the atmosphere until the trap has been removed. If the system is closed even after a brief exposure to the atmosphere, some oxygen may have already condensed.

8.3.5. Assembly of vacuum apparatus

Vacuum apparatus must be assembled in a way to avoid strain. With properly assembled joints, sections of the apparatus can be moved without transmitting strain to the necks of the flasks or joints. Heavy apparatus should be supported from below as well as by the neck.

Vacuum apparatus should be placed well back on the bench or in the hood where it will not be easily struck by personnel or the hood doors.

8.4. SAFE HANDLING OF CERTAIN VENDOR-SUPPLIED COMPRESSED GASES

Handling advice is provided here for 12 gases chosen because they are particularly common in laboratories, and/or are potentially dangerous.

8.4.1. Acetylene

Acetylene has the highest positive free energy of formation of any compound that most chemists ever encounter, and hence is the most thermodynamically unstable common substance. Moreover, it has an exceedingly wide explosive range (2-80%), and it can deflagrate in the absence of air to give oligomers or polymers. Its stability is markedly enhanced by the presence of small amounts of other compounds such as methane. Acetylene from a cylinder is safer to handle because it is dissolved in acetone in the cylinder. For some uses, such as the preparation of acetylides, it is necessary to scrub the gas to remove the acetone. Such purified acetylene is far more dangerous than acetylene from the cylinder. The stability of pure acetylene is related to the diameter of the pipe used to transport it, being less stable in wide-bore piping. The handling and use of acetylene under pressure is extremely hazardous. In the absence of compelling reasons to the contrary, all reactions and operations involving acetylene must be carried out in a pressure laboratory that has the necessary facilities, as well as expertise and experience, for its safe handling.

Handling procedures

The lowest acetylene pressure required should always be used. Under no circumstances should acetylene be used under pressure in unprotected equipment.

A pressure of 103 kPa is recognized as the maximum allowable for supply lines and regulator systems. However, even below this pressure a serious hazard exists, particularly in closed-systems containing more than 1 dm^3 of gaseous acetylene.

Equipment

Only equipment approved by recognized safety agencies should be used. Acetylene reacts with copper, silver, or lead to form the corresponding metal acetylides, which are shock-sensitive and explosive. Alloys of these metals, including solders, should not be used in contact with acetylene unless they have been specifically approved for this purpose. If it is known or suspected that acetylides have been formed, a supervisor or safety officer should be consulted as for explosive compounds in general (Chapter 11).

Contaminated piping should not be used. Acetylene reacts violently with oxidizing agents such as chlorine or oxygen. Explosive decomposition is known to be initiated by a variety of conditions, particularly elevated temperatures. When an acetylene cylinder is connected to a pressure reactor, a check valve with an appropriate valving system should be used to prevent flashback into the supply system.

Only pressure regulators approved for use with acetylene should be used. These are fitted with a flame arrestor. All repairs or modifications of acetylene regulators should be done by qualified personnel.

Previously used gauges may be contaminated and should not be fitted to acetylene cylinders except after thorough reconditioning and inspection by qualified personnel. Only gauges that have Bourdon tubes constructed of stainless steel or an alloy containing less than 60% copper should be used. Ordinary gauges usually contain brass and bronze parts that can lead to acetylide formation.

The acetylene in cylinders is in solution under pressure in a porous, acetone-impregnated, monolithic filler and is safe to handle only in this state. Such a cylinder must always be positioned vertically with the valve end up when gas is withdrawn from it. Acetylene cylinders must be protected from mechanical shock.

Purification

The purification of acetylene at atmospheric pressure and room temperature is most efficiently achieved by scrubbing the gas successively through concentrated sulfuric acid and caustic traps. A suitable grade of activated alumina, an all-purpose solid absorbent, is recommended where purification over a solid is desired. Activated carbon must be avoided because the heat of absorption may be sufficient to trigger thermal decomposition of the acetylene.

8.4.2. Anhydrous ammonia

A direct flame or steam jet must never be applied to a cylinder of ammonia. If it becomes necessary to increase the pressure in a cylinder to promote more rapid discharge, the cylinder should be moved into a warm room. Extreme care should be exercised to prevent the temperature from rising above 50 $^\circ$C.

Only steel valves and fittings should be used on ammonia containers. No copper, silver, zinc, or their alloys should be permitted to come into contact with ammonia.

Respiratory protective equipment of a type approved by regulatory agencies for anhydrous ammonia should always be readily available in places where the gas is used. Proper protection should be given to the eyes by the use of goggles or large-lens safety glasses to eliminate the possibility of liquid ammonia coming in contact with the eyes and causing injury. The threshold limit value (TLV) of ammonia is 25 cm^3/m^3.

Ammonia leaks can be detected with sulfur tapers or sensitive papers. Both of these items and instructions for their use may be obtained from the cylinder supplier.

8.4.3. Boron(III) fluoride
Toxicity

At high concentrations, boron(III) fluoride causes burns to the skin similar to those caused by hydrogen fluoride, although the burns do not penetrate as deeply as those of hydrogen fluoride (Section 6.8.2). The TLV of boron(III) fluoride is 1 cm^3/m^3.

Handling procedures

Boron(III) fluoride, b.p. -100 °C, forms dense white fumes in contact with the atmosphere. Even after a cylinder valve has been tightly closed, the fumes will linger around the outlet for as much as 0.5 hour, and frequently cause the user to believe that the valve itself is leaking. In addition, the gas is inherently difficult to control through valves and piping, and even the best of equipment is apt to show slight signs of leaking, which will cause an abundance of fumes.

It is essential when using boron(III) fluoride to have a trap in the delivery tube to prevent impurities from being sucked back into the cylinder. Certain chemicals such as water, if drawn back into a boron(III) fluoride cylinder, can build up pressure that may cause the cylinder to burst.

Every boron(III) fluoride valve is equipped with a device consisting of a platinum disk in the back of a plug containing a metal that will melt at approximately 70 °C. Frequently, a similar safety device is inserted in the base of the cylinder.

For disposal of excess boron(III) fluoride see Section 10.2.3.

8.4.4. Carbon monoxide
Toxicity

Carbon monoxide is a direct and cumulative poison with a TLV of 50 cm^3/m^3. It combines with haemoglobin of the blood to form a relatively stable compound, carboxyhaemoglobin, which is not an oxygen carrier. Fatality may occur when approximately one-third of the haemoglobin has combined in this way. The gas is odourless and insidious. Exposure to carbon monoxide in air 1500-2000 cm^3/m^3 for 1 hour is dangerous, and exposure to 4000 cm^3/m^3 may be fatal in less than 1 hour. Headache and dizziness are the usual symptoms of carbon monoxide poisoning, but occasionally the first evidence of poisoning is collapse.

Handling procedures

Carbon monoxide should be used only in areas that have adequate ventilation at all times. A trap or vacuum break should always be used to prevent impurities from being sucked back into the cylinder.

8.4.5. Chlorine

Toxicity

Chlorine, b.p. -34 °C, has a penetrating, irritating odour detectable at about 0.3 cm^3/m^3. Minimal irritation of the throat and nose is noticed at about 2.6 cm^3/m^3 and painful irritation at about 3.0 cm^3/m^3. Exposure to about 17 cm^3/m^3 causes coughing, and levels as low as 10 cm^3/m^3 may cause lung oedema. Human exposure to 14-21 cm^3/m^3 for 30 minutes to 1 hour is regarded as dangerous and may, after a delay of 6 or more hours, result in death from anoxia due to serious pulmonary oedema. The TLV of chlorine is 1 cm^3/m^3.

Handling procedures

Wherever chlorine is used, respiratory protective equipment should be available. Proper protection should be afforded to the eyes by the use of goggles or large-lens spectacles to eliminate the possibility of splashing from liquid chlorine. Chlorine should be used only by experienced or properly instructed persons.

Only auxiliary valves and gauges designed solely for chlorine should be used. Stainless steel equipment should not be used. Every precaution must be taken to avoid drawing liquids back into chlorine containers. Accordingly, there should be a trap next to the valve, and the valve should be closed immediately after the container has been emptied.

Chlorine leaks may be detected by passing a rag dampened with aqueous ammonia over the suspected valve or fitting. White fumes indicate escaping chlorine gas.

Chlorine should be kept away from easily oxidized materials. It reacts readily with many organic chemicals, sometimes explosively. Because of the high toxicity of chlorine, laboratory operations using it should be carried out in a hood, and rubber or neoprene gloves should be worn.

8.4.6. Fluorine

Toxicity

Fluorine causes deep penetrating burns on contact with the body, an effect that may be delayed and progressive, as are burns by hydrogen fluoride (Section 6.8.2). The hazard of exposure to fluorine in the atmosphere is at least as great as that of chlorine. The TLV of fluorine is 1 cm^3/m^3.

Other properties

Fluorine reacts vigorously with most oxidizable substances at room temperature, frequently with immediate ignition, and with most metals at elevated temperatures. In addition, it reacts vigorously with silicon-containing compounds and can thus support the continued combustion of glass and asbestos. Fluorine forms explosive mixtures with water vapour, ammonia, hydrogen, and most organic vapours. It is strongly recommended that work with fluorine be conducted with a compressed gas mixture such as fluorine/nitrogen (25:75). Such mixtures are available from compressed gas vendors, and are safer than pure fluorine.

Handling procedures

Because of the high activity of fluorine, the area in the vicinity of a fluorine cylinder and its associated apparatus should be well ventilated and cleared of easily combustible material.

When a cylinder of fluorine is to be opened, the user should be protected by a suitable shield (Section 7.2.3) and the valve should be opened by remote control. Any apparatus that is to contain fluorine under pressure should be surrounded by a protective barrier. Fluorine cylinder valves are not adapted to fine adjustment, and the flow of fluorine from a cylinder should therefore be controlled by a needle valve located close to the cylinder and operated by remote control. All equipment that may be in contact with fluorine should be completely dry.

Protective measures against fluorine are not fully developed, and entry into contaminated zones should be avoided. Only respiratory protective equipment supplied with a positive-pressure atmosphere (Section 2.4.2) is advised. If concentrations over 2 cm^3/m^3 are reached, a pressure-demand self-contained breathing apparatus or a positive pressure air line respirator that has escape-cylinder provisions must be used. Gauntlet type rubber gloves, rubber aprons, and face shields give only temporary protection against fluorine and, if brought into local contact with a high concentration of fluorine from leak, may inflame. A thorough flushing of fluorine lines with an inert gas should precede any opening of the lines for any reason.

The reaction of fluorine with some metals is slow because of the formation of a protective metallic fluoride film. Brass, iron, aluminium, and copper, as well as certain of their alloys, react in this way at ambient temperatures and atmospheric pressure. Thus, these metals can be rendered inert by passing fluorine gas highly diluted with argon, neon, or nitrogen through tubing or over the surface with proper precautions and gradually increasing the concentration of fluorine. Protected apparatus is safe to use only if the protective coating is not cracked or dislodged to create a "hot-spot."

Once a fire that involves fluorine as an oxidizer has started, there is no effective way of stopping it other than shutting off the source of fluorine. The area should be cleared and the fire allowed to burn itself out. No attempts should be made to extinguish it with water or chemicals. Anyone in the vicinity of such a fire should wear impervious clothing and supplied-air or self-contained breathing apparatus.

8.4.7. Hydrogen

Hydrogen unlike other gases, shows a temperature increase when the gas is expanded at a temperature higher than its inversion point (-68.6 °C). This is known as the inverse Joule-Thomson effect. A cylinder of hydrogen will sometimes emit a flash of fire when the cylinder valve is opened suddenly, permitting a rapid escape of gas. It is thought by some that the inverse Joule-Thomson effect, plus the static charge generated by the escaping gas, may cause its ignition.

Hydrogen has an extremely wide flammability range, the highest burning velocity of any gas and, although its ignition temperature is reasonably high, a very low ignition energy. It burns with a non-luminous flame that is often invisible in daylight. Hydrogen is not intrinsically toxic.

Handling procedures

Hydrogen presents both explosion and fire hazards when released. However, although its wide range of flammability and high burning rate accentuate these hazards, its low ignition energy, low heat of combustion on a volume basis, and non-luminous (low thermal radiation level) flame are counteracting effects.

As a result of its low ignition energy, small sources of heat (e.g., friction and static-generation) may cause ignition when the gas is released at high pressure. Accordingly, hydrogen is frequently thought of as "self-igniting". When hydrogen is released at low pressures however, self-ignition is unlikely. Hydrogen combustion explosions are characterized by very rapid pressure increases that are extremely difficult to vent effectively. Open air explosions have occurred from large releases of gaseous hydrogen.

Because of its very low boiling point, contact between liquid hydrogen and air can result in condensation of air including its oxygen and nitrogen components. A mixture of liquid hydrogen and liquid oxygen is potentially explosive. Accidents from this source have generally been restricted to small containers of liquid hydrogen open to the atmosphere.

Because of the low density of hydrogen, it has a high diffusion rate in air, which makes it difficult for hydrogen to accumulate in conventional structures and tends to reduce its combustion explosion hazard.

Escaping gaseous hydrogen seldom presents an emergency other than fire or explosion because either it is ignited promptly or rises rapidly in the atmosphere. Hydrogen gas vaporizing from the liquid phase near its normal boiling point is slightly heavier than air at 20 °C, and this causes it, together with the visible fog of condensed water vapour so created, to spread along the ground for considerable distances.

Water should be applied to containers of hydrogen exposed to fire, and the flow of gas should be stopped if possible. Because a hydrogen flame is often invisible in daylight and produces low levels of thermal radiation, it is possible to walk into the flame inadvertently. Thus great care should be exercised when approaching a hydrogen fire.

8.4.8. Hydrogen bromide and hydrogen chloride

Hydrogen bromide and hydrogen chloride are corrosive gases that have pungent, irritating odours. Although both are colourless, they fume in moist air. In the cylinder under pressure, both exist as gases over a liquid phase. Under such conditions, the cylinder pressure is equal to the vapour pressure of the liquid; at 25 °C, this is 4.22 MPa for hydrogen chloride and 2.2 MPa for hydrogen bromide. Although neither hydrogen chloride nor hydrogen bromide is combustible, both react with common metals to produce hydrogen, which may form explosive mixtures with air.

Toxicity

Both hydrogen chloride and hydrogen bromide are highly toxic gases and severely irritate the upper respiratory tract. They can cause death as a result of oedema or spasm of the larynx and inflammation of the upper respiratory system. Concentrations of 0.13-0.2% (v/v) are lethal for man within a few minutes, but their irritating odour provides adequate warning for prompt withdrawal from a contaminated area. They are, however, also corrosive to the skin and eyes and can cause severe burns. The TLV of hydrogen chloride is 5 cm^3/m^3; of hydrogen bromide 3 cm^3/m^3.

Handling procedures

These gases should be handled only in adequately ventilated areas. A check valve, vacuum break, or trap should always be used to prevent foreign materials from being sucked back into the cylinder as this can cause the development of dangerous pressures. Laboratory workers who handle either of these gases should wear protective apparel, including rubber gloves, suitable gas-tight chemical safety goggles, and clothing such as a rubber or plastic apron (Sections 2.2, 2.3). Proper respiratory equipment should be available (Section 2.4.2).

Significant leaks of hydrogen chloride or hydrogen bromide will be evident by the formation of dense white fumes on contact with the atmosphere. Small leaks can be detected using a small beaker of concentrated ammonium hydroxide, the formation of dense white fumes confirming the existence of a leak.

8.4.9. Hydrogen sulfide

Toxicity

Hydrogen sulfide, b.p. -62 °C, is extremely dangerous. Human exposure to relatively low concentrations has caused corneal damage, headache, sleep disturbances, nausea, weight loss, and symptoms suggestive of brain damage. Although its rotten egg smell is strong and characteristic, initial exposure to a relatively low concentration can deaden the olfactory nerves of most people so that further exposure is not detected and is extremely dangerous. Higher concentrations can cause irritation of the lungs and respiratory passages and even pulmonary oedema. Exposure to 210 cm^3/m^3 for 20 minutes has caused unconsciousness, arm cramps, and low blood pressure. Coma may occur within seconds after one or two breaths at high concentrations and be followed rapidly by death. For example, workers have died after exposure to 930 cm^3/m^3 for less than 1 minute. The TLV is 10 cm^3/m^3.

Handling procedures

Respiratory protective equipment (Section 2.4.2) should always be readily available in places where this gas is used. A gas mask should be used only when the concentration of hydrogen sulfide is low. Otherwise, a supplied-air respirator or self-contained breathing apparatus is required.

Hydrogen sulfide should never be used from a cylinder without reducing the pressure through a regulator attached directly to the cylinder. Cylinders should not be stored in small, unventilated rooms, as death has resulted from entering such a room containing a leaking cylinder.

8.4.10. Oxygen

Oxygen and organic oils combine explosively. Therefore valves, regulators, gauges, and fittings used in oxygen supplies must not be lubricated with oil or any other combustible substance. Molybdenum(IV) sulfide powder or a fluorolube are acceptable. Oxygen cylinders or apparatus should not come in contact with easily combustible materials or be stored near them, and they should not be handled with oily hands or gloves. All piping and apparatus must be cleaned meticulously to remove dirt and organic residues before coming into contact with oxygen. Even then, the oxygen should be introduced cautiously, and pressure developed slowly. Oxygen regulators, hoses, and other appliances should not be interchanged with similar equipment intended for use with other gases. Oxygen is not flammable, but supports combustion. Once a fire begins in pure oxygen, almost anything, including many metals, will burn.

8.4.11. Phosgene

Toxicity

Phosgene, b.p. 8.3 °C, has a geranium-like odour at low levels. The symptoms of overexposure are dryness or a burning sensation in the throat, numbness, vomiting, and bronchitis. An airborne concentration of 5 cm^3/m^3 may cause eye irritation and coughing in a few minutes. It can cause severe lung injury in 1-2 minutes at a level of 20 cm^3/m^3. Exposure to concentrations above 50 cm^3/m^3 is likely to be fatal. The TLV of phosgene is 0.1 cm^3/m^3.

Handling procedures

No person should work with this compound without being fully familiar with its toxic effects and proper handling procedures. Ventilation is extremely important, and respiratory protective equipment should be available (Section 2.4.2). Corrosion problems are not serious, and brass fittings may be used.

In case of a leak in a phosgene cylinder, the brass cylinder cap should be affixed as tightly as possible and the cylinder placed in the coolest ventilated spot that is available. The manufacturer should be notified at once.

Small amounts of phosgene can be disposed of by alkaline hydrolysis (Section 10.1.10).

Part III

Safe Storage and Disposal of Waste Chemicals

9

Recovery, Recycling and Re-use of Laboratory Chemicals

To the extent that chemicals can be recovered, recycled, or re-used safely at net costs less than the costs of disposal as waste, there is an economic incentive to do so. In addition, materials that are recovered, recycled, or re-used do not become a problem in the environment.

Even though recovery, recycling, and re-use of laboratory chemicals may not be an economical option currently for a given laboratory, it is an option that should be continually assessed. Disposal of hazardous waste in secure landfills is becoming increasingly problematic, and the capacities of such landfill facilities are quite limited. Moreover, the prospects for easier access to incinerators that will accept small volumes of chemically diverse laboratory wastes are questionable. These factors may change the relative economics of recovery, recycling, and re-use. Recovery of laboratory chemicals can be a valuable educational exercise in academic laboratories when carried out under proper supervision.

9.1. RECOVERY OF VALUABLE METALS

The metal content of some materials used in laboratories has sufficient value to make recovery economically feasible. Small amounts of valuable metals can sometimes be recovered and recycled in the laboratory. Such recovery procedures should be carried out by, or under the direct supervision of, a trained professional who understands the chemistry involved. Larger quantities of some metals or metal compounds can often be sold to suppliers or reprocessors for recovery. The original supplier should be able to furnish the names and locations of reprocessors.

9.1.1. Mercury

Metallic mercury should be collected for recovery and recycling.

Manipulations should be carried out in a hood as far as possible. Small quantities can be made relatively free of insoluble contaminants by repeated filtering through a conical filter paper with a small hole at the apex. Insoluble contaminants float to the top and collect on the sides of the filter cone. The filter paper should be discarded for disposal in a secure landfill. Liquid mercury in quantities greater than 5 kg can be sold to a commercial reprocessor. Smaller quantities may be accepted at no charge if the donor pays transportation costs. Large quantities can be sent in standard 35-kg flasks to reprocessors, who will usually supply the flasks. Most reprocessors will purchase mercury only from institutions, not from individuals.

Much of the metallic mercury from a spillage can be collected in a bottle equipped with an eyedropper type nozzle and connected to a vacuum aspirator. Small droplets can be amalgamated with zinc dust and the resulting solid swept up. Droplets in floor crevices from spills, or in crevices in metal containers, can be converted to mercury(II) sulfide by dusting with sulfur powder. There are commercial kits that contain materials and equipment for cleaning up minor spills. Mercury can be recovered from its salts by converting them to the sulfide. Amalgams and mercury(II) sulfide are accepted by reprocessors for recovery, in sufficiently large quantities.

9.1.2. Silver

A procedure developed for the recovery of silver (and mercury) salts from solutions remaining from chemical oxygen demand tests can be used to recover metallic silver from aqueous solutions of its salts.

$$Ag^+ + NaCl \longrightarrow AgCl + Na^+$$
$$AgCl + Zn + H_2SO_4 \longrightarrow Ag + H_2 + \text{soluble Zn salts}$$

The solution of a silver salt is acidified with 6 mol/dm^3 nitric(V) acid (pH 2) and treated with a 10% excess of 20% (w/v) aqueous sodium chloride solution. The precipitated silver chloride is collected in a Buchner funnel and washed twice with warm 2 mol/dm^3 sulfuric acid and twice with distilled water. The silver chloride is dried and ground to a fine powder. One hundred grams of the dry silver chloride powder are mixed thoroughly with 50 g of pure granulated zinc metal and the mixture is stirred with 500 cm^3 of 2 mol/dm^3 sulfuric acid. *(CAUTION: This operation should be performed in a hood away from any source of ignition because hydrogen is evolved.)* When the zinc has dissolved, the supernatant solution is decanted, and the crude silver is again mixed with 50 g of granulated zinc and treated with 500 cm^3 of 2 mol/dm^3 sulfuric acid. After the zinc has dissolved, about 5 cm^3 of concentrated sulfuric acid are added carefully, and the mixture heated to 90 °C with stirring for a few minutes. The silver is separated by filtration and washed with distilled water until the washings are free of sulfate (BaCl$_2$ test). A sample of the resulting silver should give a clear solution in concentrated nitric(V) acid. If the solution is turbid, indicating the presence of silver chloride, the treatment with zinc and sulfuric acid should be repeated.

This procedure should give silver of about 99.9% purity. The zinc salt solutions can be flushed down the drain if local regulations permit, or the zinc can be precipitated as the carbonate for landfill disposal.

Scrap silver-based photographic film and photographic fixer can be recovered by commercial reprocessors. It is possible to purchase electrolytic units that are designed for the recovery of silver from photographic solutions. Silver can also be recovered from photographic fixer solutions by passing them through a cartridge (commercially available) filled with steel wool or iron particles. Metal exchange occurs, and the silver drops as a sludge through a screen at the bottom of the cartridge into a liquid-filled space in the cartridge container. The dilute iron-containing effluent can be poured down the drain. This procedure is low-cost and reduces the silver content of a solution to 20 mg/dm^3 or less; it works more efficiently as a continuous rather than an intermittent process.

9.1.3. Other valuable metals

It is worth recovering noble metals such as platinum, palladium, rhodium, iridium or ruthenium from spent catalysts. Generally, the original supplier will accept quantities greater than a specified minimum for recovery. The supplier should be consulted concerning any pretreatment that may be required.

Small quantities can be recovered in the laboratory by chemical procedures. However, the chemistry of these metals is sufficiently distinct that no general procedure is applicable, and specialized procedures are needed.

9.2. RECOVERY OF SOLVENTS BY DISTILLATION

If the work in a laboratory is such that waste solvents of known composition are produced regularly, then recovery may be economically feasible. The costs of recovery should be balanced against the combined costs of purchasing new solvents and of disposing of the used ones. An assessment of solvent use can reveal those that are used in sufficient quantity to warrant segregation of used material for recovery. Some commercial firms recover solvents in the quantities that are used in a moderate-sized laboratory.

It may be possible to recover common organic solvents by distillation in laboratory bench-top stills operated by students or technicians under proper supervision. Some equipment suppliers sell laboratory-size stills designed for solvent recovery. Larger recovery operations may be conducted in commercial solvent stills. Special care must be exercised in distillation recovery of solvents that boil near room temperature to ensure that the condenser and its coolant are capable of preventing significant quantities of uncondensed vapour from escaping into the atmosphere. Peroxide-forming solvents such as ethers should be excluded from any distillation recovery operation.

Effective distillation and recovery is determined by proper segregation of the used solvents at the point of generation. Each solvent should be kept separate, and solvents for recovery should not be highly contaminated. The laboratory waste-disposal plan should identify those solvents that are to be recovered and should define the markings on the containers that are to be used for their collection.

A potential problem is the acceptability of the recovered solvent to chemists or other users who may be concerned about purity. This problem may be overcome by a pilot trial with the recovered solvent. With proper segregation at the point of generation and careful distillation procedures, solvents of almost any desired degree of purity can be produced. Recovered solvents of potentially low purity may be used as thinners, solvents, and degreasers by other academic or operational departments of the institution, provided that it is first determined by analysis or from knowledge of the source that the solvent does not contain hazardous contaminants.

9.3. EXCHANGE OF UN-NEEDED CHEMICALS

Chemicals that are no longer needed by one laboratory may be of use to another. Some laboratories have set up a retained chemical store. Partially used reagents in their original containers are sent to this store, indexed by card file or computer, and stored on indexed shelves for easy retrieval. Laboratory personnel who need a certain reagent should check the store before placing a purchase order. Some institutions accept compounds that have been synthesized in their own laboratories with the requirement that the sample is properly labelled with chemical identity, name of the worker who synthesized it, date, and notebook reference.

It is essential that a retained chemical store be well ventilated and equipped with a sprinkler system. Incompatible chemicals (Appendix E) should be kept apart. The area should be inspected regularly. Samples that show evidence of deterioration or on which labels have become illegible should be discarded. Some samples can be retained by relabelling. If the index is computerized, and the number of chemicals is not too great, a list can be distributed periodically to all units of the organization. The retained chemical store can also be a clearing house for standing requests for excess chemicals. Some institutions have found that they receive requests for chemicals from one laboratory that have been returned from another.

Arrangements may also be possible for exchange of un-needed chemicals between different institutions. For example, exchange, donation, or sale at a nominal price of unopened containers might be arranged between laboratories that are located near each other or among academic institutions that operate under a common administration. The possibility of disposing of un-needed chemicals through a regional chemical exchange should also be considered.

Industrial laboratories sometimes donate such chemicals to nearby academic institutions, and this practice is to be encouraged. However, to avoid loading the recipient with unwanted chemicals, a member of the academic staff should be responsible for selecting those reagents which the institution can use.

It should be possible to exchange or donate unopened or partially used samples of chemicals that are in their original containers, provided that:

1. The donor has a control system that gives reasonable assurance of the identity of the chemical.
2. The recipient accepts full responsibility for the identity and quality of such chemicals.

A retained chemical store or exchange should not accept chemicals that are capable of forming dangerous peroxides (Appendix B, Table B.1 and B.2) unless they are known to contain inhibitors.

9.4. USE OF UN-NEEDED CHEMICALS AS FUEL

The heat of combustion of certain types of organic chemicals can be partially recovered by adding them in small quantities to the fuel feed of the institution's power-generating unit. In general, this procedure can be applied to hydrocarbons and oxygen- or nitrogen-containing compounds. Compounds that contain halogen or sulfur should not be burned, except as minor contaminants of an acceptable chemical type. Highly toxic compounds, explosives, compounds with a low flash point, and organometallics must be excluded from this process. Procedures for this method of disposal are outlined in Chapter 16.

10

Procedures for Laboratory Destruction of Hazardous Chemicals

It is possible to reduce or destroy the dangerous nature of many hazardous chemicals by chemical reaction in the laboratory. Although "in-house" chemical destruction is not likely to be an economically practical solution to the general problem of disposal of laboratory wastes, many laboratories may find it useful for certain wastes. Academic laboratories may find that the destruction of some chemicals can be made an effective part of their instructional programme, at the same time reducing the quantities of wastes that have to be disposed of by other means.

This chapter addresses the problem by giving guidelines and procedures for destroying and disposing of common classes of laboratory chemicals.

Several destruction procedures are presented in the detail that is usually reserved for synthetic procedures. It would be helpful if chemists who work out details of destruction procedures were to write them up in the style of *Organic Syntheses*, have them checked experimentally in another laboratory for efficacy and safety, and then publish them. For example, the *International Agency for Research on Cancer* is publishing a series of monographs on procedures for destroying carcinogens including aflatoxins, N-nitrosamines, polycyclic aromatic hydrocarbons, hydrazines, N-nitrosamides, haloethers, aromatic amines, and antineoplastic agents. Clearly, the hazards presented by these compounds justify the costs of destroying them. It is a fundamental principle of good laboratory practice that no worker should undertake manipulations with any chemical without first understanding its properties and possible hazards. The destruction/disposal procedures presented here are intended to be carried out only by, or under the direct supervision of, a trained scientist or technologist who understands the chemistry involved. The laboratory practices recommended in Chapters 1, 2 and 6-8 should be followed, making proper use of laboratory fume hoods (Chapter 3). The procedures are intended for application to laboratory quantities, i.e. not more than a few hundred grams. Because hazards tend to increase exponentially with scale, larger quantities should be treated only in small batches unless a qualified chemist in the laboratory concerned has demonstrated that the procedure can be scaled up safely. Some laboratories have prepared handbooks on the properties, health hazards, and methods of disposal of the compounds that they use.

10.1. ORGANIC CHEMICALS

Most organic compounds can be destroyed in properly designed and operated incinerators (Chapter 16), and the chemical classes that are suitable for this method of destruction are indicated in the following sections. However, many laboratories do not have their own incinerator and find that commercial incinerator operators either are not willing to accept laboratory waste or are too expensive to employ. Accordingly, they may be obliged to dispose of their hazardous waste in a secure landfill (Chapter 18).

Some chemicals can be disposed of in a sanitary sewage system (Chapter 15) or by chemical destruction in the laboratory. These options will be considered in the following discussion of the principal classes of organic chemicals.

10.1.1. Hydrocarbons

This class includes alkanes, alkenes, alkynes, and arenes. They burn well and can be disposed of by incineration or as fuel supplements (Chapter 16). Many of those commonly used in laboratories are easily ignited. They can be put into a secure landfill only in small quantities, packaged in a lab pack (Chapter 18). Because they are virtually insoluble in water and quite volatile, only traces should be put down the drain.

Some alkenes, especially cyclic ones like cyclohexene, may form explosive peroxides on long storage with access to air. Old samples should be tested for peroxides and treated appropriately if peroxides are present (Section 10.1.16).

Some polycyclic arenes, such as benzo[a]pyrene and 1,2-benzanthracene, are potent animal carcinogens and suspected human carcinogens. Although such hydrocarbons can be incinerated or put into a lab pack, they, or mixtures containing them, must be packaged separately and carefully labelled so that operators can handle them without being exposed. Up to a few grams of a polycyclic arene can be destroyed by oxidation in a freshly prepared solution of potassium manganate(VII) in 3 mol/dm^3 sulfuric acid.

Procedure for destroying one gram of a polycyclic arene
Potassium manganate(VII) (94.8 g, 0.60 mol) is dissolved in 2 dm^3 of 3 mol/dm^3 sulfuric acid by gradually adding it to the stirred acid with sufficient cooling to keep the mixture near room temperature. A solution of 1.0 g of polycyclic arene in 400 cm^3 of acetone is gradually added to the fresh manganate(VII) solution at room temperature with stirring. The mixture should stand at room temperature for at least 1 hour. If the purple colour disappears during this period, more manganate(VII) solution is added. Enough solid sodium hydrogen sulphate(IV) is added with stirring to reduce excess manganate(VII) (purple colour disappears) and dissolve the manganese(IV) oxide. The mixture is neutralized (pH paper) with 10 mol/dm^3 sodium hydroxide and flushed down the drain. This is a potentially dangerous reaction and the method should not be scaled up to destroy larger quantities.

Glassware is decontaminated by being rinsed several times with acetone. The combined rinsings are treated as described above.

10.1.2. Halogenated hydrocarbons

Several halogenated hydrocarbons are common laboratory solvents, for example methylene chloride, chloroform, carbon tetrachloride, tetrachloroethylene, 1,1,1-trichloroethane, 1,1,2-trichloro-1,2,2-trifluoroethane, chlorobenzene, and 1,2-dichlorobenzene. Other halogenated hydrocarbons are important laboratory reagents, examples being t-butyl chloride, benzyl chloride, ethyl bromide, methyl iodide, bromobenzene, methylene iodide, 1,2-dibromoethane, and allyl chloride. Many of these are toxic, and some are carcinogenic to laboratory animals, indicating the need for prudence in handling and disposing of them.

Those that are used as solvents may be recovered by distillation. Their low solubility in water and volatility make more than traces unacceptable in the sanitary sewer. Many laboratories dispose of irrecoverable halogenated hydrocarbons in landfill lab packs. When blended with excess non-halogenated solvents, they can be destroyed by incineration, but more than small quantities require an incinerator equipped with a scrubber to remove the hydrogen halide from combustion gases.

A few highly halogenated arenes, such as polychlorinated biphenyls, are not completely destroyed by most incinerators, and few incinerators have the demonstrated capability for the efficient destruction of these refractory substances. In addition, some highly toxic compounds such as chlorinated dibenzodioxins may be formed from precursors at temperatures ranging from 300-500 °C. Few secure landfills can accept them lawfully.

Up to a few hundred grams of alkyl halides can be destroyed by hydrolysis with ethanolic potassium hydroxide. The general procedure given below is applicable to most primary, secondary, and tertiary alkyl chlorides, bromides, and iodides. It can also be used for allyl and benzyl halides, including fluorides, and for compounds that have two or even three chlorine, bromine, or iodine atoms on one carbon. It is not useful for vinyl or aryl halides or for most alkyl fluorides because they are too unreactive. The procedure can also be used to hydrolyse the following classes of alkylating agents, all of which contain some members that are human or animal carcinogens: dialkyl sulfates, alkyl alkanesulfonates and arenesulfonates, chloromethyl ethers, epoxides, and aziridines. Although some of the more active alkylating agents can be hydrolysed effectively by treatment with aqueous sodium hydroxide (Section 10.1.10), the following procedure is more broadly applicable. It uses 20% excess potassium hydroxide based on the reaction:

$$RX + KOH \longrightarrow ROH + KX$$

An additional equivalent of potassium hydroxide must be used for each additional hydrolyzable group in the compound being treated. Competing reactions, such as the formation of ethyl ethers (ROC_2H_5) or dehydrohalogenation, also destroy the organic halide. Dehydrohalogenation becomes the principal reaction with tertiary halides, generating olefins, e.g.:

$$(CH_3)_3CCl + KOH \longrightarrow (CH_3)_2C:CH_2 + KCl + H_2O$$

(CAUTION: Reactions that may generate olefins too volatile to be trapped by a reflux condenser, such as isobutylene or 2-methyl-2-pentene, should be carried out in an efficient hood away from any source of ignition.) Although a few hydrolysis products, such as the aldehyde formed from a 1,1-dihalide, may undergo condensation reactions, this does not interfere with the destruction of the organic halide.

Procedure for hydrolyzing 1.0 mol of an alkyl halide

Place 79 g (1.20 mol) of 85% potassium hydroxide pellets in a 1-dm^3 three-necked flask equipped with a stirrer, water-cooled condenser, dropping funnel, and heating mantle or steam bath. With brisk stirring, 315 cm^3 of 95% (v/v) ethanol are added rapidly. The potassium hydroxide dissolves within a few minutes, causing the temperature of the solution to rise to about 55 °C. The solution is heated to reflux gently, and the liquid alkyl halide (or its solution in 95% ethanol if it is a solid) is added dropwise. The rate of addition and the heat input are adjusted to maintain gentle refluxing. The dropping funnel is rinsed with a little ethanol, and stirring and reflux are continued for 2 hours. Stirring is essential to prevent bumping caused by the potassium halide that is likely to precipitate. If the reaction products are water-soluble, the mixture is diluted with 300 cm^3 of water, cooled to room temperature, neutralized, and washed down the drain with about 50 volumes of water. Otherwise, the mixture is neutralized and sent to an incinerator or landfill.

The procedure is most useful for halides that are hydrolysed to water-soluble products. Those that are not can generally be incinerated or sent to a secure landfill; however, for some in the latter category, the procedure may be useful for decomposing a toxic compound to substances which are allowed in landfills. Chloromethyl ethers can be destroyed by treatment with a 25-fold excess of 6% (w/v) aqueous ammonia at room temperature for 3 hours.

Many organic halides can undergo slow spontaneous hydrolysis. If they contain moisture when packed for landfill disposal, the hydrogen halide generated by hydrolysis will slowly destroy the bottle caps unless these are made of polypropylene or lined with Teflon®.

10.1.3. Other halogenated compounds

This broad class comprises halogenated hydrocarbons that contain a functional group such as hydroxy, carboxy, thiol, amino, or nitro. Members of this class can be incinerated or, if not reactive, put into a secure landfill. Hydrolysis with ethanolic potassium hydroxide can destroy many of these compounds. Those that contain base-reactive groups will require additional potassium hydroxide. Acid halides are discussed separately (Section 10.1.10).

In general, halogen substituents decrease the water-solubility of organic compounds, but a few are sufficiently soluble to be flushed down the drain, e.g., 2-chloroethanol and chloroacetic acid. However, some that have high solubility e.g., fluoroacetic acid, are too toxic for such disposal.

10.1.4. Alcohols and phenols

Alcohols are used extensively as solvents and reagents in the laboratory. They can be incinerated, burned as fuel supplement, or put into a secure landfill. Many of the common aliphatic alcohols are readily soluble in water, have low toxicities, are easily biodegradable, and can be flushed down the drain.

Phenols tend to be more toxic. They are skin irritants and some are allergens, so care must be taken to avoid skin contact while disposing of them. They can be incinerated or put into a secure landfill. Many phenols, particularly phenol itself and its monosubstituted derivatives, are efficiently degraded by biological treatment systems and can be destroyed in an on-site biological treatment facility (Section 15.2). Phenols that enter water-treatment plants may become chlorinated to chlorophenols, which give drinking water a bad taste. Therefore phenols should never be allowed to enter storm sewers or other drains that lead directly to streams that are used as sources of drinking water. Even the disposal of phenols in waste water-treatment plants should be avoided because if the plant's efficiency is poor, it may let some phenolic material through to the chlorination system.

Aqueous solutions that contain low concentrations of phenols can be passed through a bed of activated carbon, which adsorbs the phenols. Small industrial adsorption units consist of a tank or column filled with carbon granules that can accommodate flow rates of about 0.3 dm^3/s. If such units are to be used regularly, facilities for regenerating the carbon can be added. On a smaller scale, a glass column filled with pelletized activated carbon (for example, Norit®) can be used to adsorb phenols from solutions. Suppliers of activated carbon can furnish information on the adsorptive capacity of various types of carbons for various compounds. Spent activated carbon can be incinerated or disposed of in a landfill, depending on what it has adsorbed. The effluent solution should contain so little phenolic material that it can be disposed of in the sanitary sewer system.

Small amounts of phenols can be destroyed in the laboratory by hydrogen peroxide in the presence of an iron catalyst: the procedure is illustrated for phenol itself.

Procedure for decomposing 0.5 mol of phenol

A solution of 47 g (0.50 mol) of phenol in 750 cm^3 of water is prepared in a 2-dm^3 three-necked flask equipped with a stirrer, dropping funnel, and thermometer. *(CAUTION: The order of addition of the reagents is important. If hydrogen peroxide and iron(II) sulfate are premixed, a violent reaction may occur.)* Iron(II) sulfate heptahydrate (23.5 g, 0.085 mol) is then dissolved in the mixture, and the pH is adjusted to 5-6 (pH paper) with dilute sulfuric acid. Then 410 cm^3 (4.0 mol) 30% (v/v) hydrogen peroxide are added dropwise with stirring over the course of about an hour. Heat is evolved, and the reaction temperature is maintained at 50-60 °C by adjusting the rate of addition and by using an ice bath if necessary. Stirring is continued for 2 hours while the temperature gradually falls to ambient. The solution is allowed to stand overnight and is then washed down the drain with a large volume of water.

This procedure can be applied to many phenols, including the cresols, monochlorophenols, and naphthols. More water is generally needed to dissolve these; thus, 0.5 mol of a cresol requires about 2.5 dm^3 of water.

10.1.5. Ethers

Open-chain monoethers, both aliphatic and aromatic, are relatively non-toxic and usually their only dangerous property is their ability to be ignited. Accordingly, they are not candidates for laboratory destruction. Except for diethyl ether, they are not sufficiently water-soluble for drain disposal. Even diethyl ether should not be put down the drain because of its volatility and flammability. Small quantities can be evaporated in a hood (with no open flame, hot surface, or other source of

168   *Chemical Safety Matters*

ignition in the hood) if the ether is free of peroxides or contains an inhibitor, which is usual with most commercial preparations. Diethyl ether can also be mixed with at least 10 volumes of higher-boiling solvents for incineration.

The common cyclic ethers tetrahydrofuran and dioxan are sufficiently water-soluble for drain disposal, although this may be prohibited in some localities because of their low flash points. They can be incinerated or put into a secure landfill.

Lower alkyl ethers and ether acetates of ethylene glycol and diethylene glycol (glyme, diglyme, carbitols, cellosolves) are water-soluble and can be disposed of down the drain with a large volume of water.

Many ethers on exposure to air can form explosive peroxides (Appendix B), the disposal of which is discussed in Section 10.1.16.

Some of the macrocyclic polyethers, "crown ethers," are quite toxic. Small quantities can be destroyed by acidic potassium manganate(VII) (Section 10.1.1).

Epoxides are potent alkylating agents, and many are highly toxic. The epoxide group can be destroyed by alkaline hydrolysis as described for alkyl halides (Section 10.1.2).

10.1.6.   Thiols

Thiols can be incinerated (provided the effluent gas is scrubbed) or put into a secure landfill. The relatively small quantities usually employed in laboratory work can be destroyed by oxidizing the thiol group to a sulfonic acid group with sodium chlorate(I). If the thiol contains other groups that can be oxidized by chlorate(I), the quantity of this reagent used must be increased accordingly.

$$RSH + 3OCl^- \longrightarrow RSO_3H + 3Cl^-$$

Procedure for oxidizing 0.5 mol of a liquid thiol

Pour 2.5 dm^3 (1.88 mol, 25% excess) of commercial laundry bleach (e.g., Clorox®, 5.25% sodium chlorate(I)) into a 5-dm^3 three-necked flask located in a fume hood. The flask is equipped with a stirrer, thermometer, and dropping funnel. The thiol (0.50 mol) is added dropwise to the stirred chlorate(I) solution, initially at room temperature. A solid thiol can be added gradually through a neck of the flask or can be dissolved in tetrahydrofuran and the solution added to the chlorate(I). Traces of thiol can be rinsed from the reagent bottle and dropping funnel with tetrahydrofuran and the rinsings added to the oxidizing solution. Oxidation usually starts soon, accompanied by a rise in temperature and dissolution of the thiol. If the reaction has not started spontaneously after about 10% of the thiol has been added, addition is stopped and the mixture warmed to about 50 °C to initiate this reaction. Addition is resumed only after it is clear that oxidation is occurring. The temperature is maintained at 45-50

°C by adjusting the rate of addition and using an ice bath if necessary. Addition requires about an hour. Stirring is continued for 2 hours while the temperature gradually falls to room temperature. The mixture should be a clear solution, perhaps containing traces of oily by-products. The reaction mixture can be flushed down the drain with excess water. The unreacted laundry bleach need not be decomposed. It is important to realize that sodium chlorate(I) solutions deteriorate on storage, and before use in degradation reactions, the strength of the active chlorate(I) content should be checked.

Calcium chlorate(I) may be used as an alternative to sodium chlorate(I) and requires only one-third as much water. For 0.5 mol of thiol, 210 g (25% excess) of 65% calcium chlorate(I) (technical grade) is stirred into 1 dm^3 of water at room temperature. The chlorate(I) soon dissolves, and the thiol is then added as in the above procedure.

Laboratory glassware, hands, and clothing contaminated with thiols can be deodorized by a solution of Diaperene®, a tetraalkylammonium salt used to deodorize containers in which soiled diapers have been washed.

10.1.7. Other organosulfur compounds

Sulfonic acids are discussed in Section 10.1.9, and sulfonyl halides in Section 10.1.10. Sulfides, sulfoxides, sulfones, thioamides, and sulfur heterocycles can be incinerated or landfilled. Small amounts of sulfides, RSR′, can be oxidized to sulfones, RSO$_2$R′, to eliminate their disagreeable odours. The chlorate(I) procedure used for thiols (Section 10.1.6) can be employed for this purpose, although the resulting sulfones are often water-insoluble and may have to be separated from the reaction mixture by filtration.

Carbon disulfide poses a special problem. It is highly volatile (b.p. 46 °C), highly toxic, and readily forms ignitable or explosive mixtures with air; even a hot steam line can ignite such a mixture. Incineration is possible, but must be carried out with great care. Containers of more than approximately 100 g of carbon disulfide can cause explosions in incinerators. It should be diluted with 10 volumes of a higher-boiling solvent before incineration. Its low boiling point, toxicity, and odour preclude landfilling. Small quantities should be destroyed by chlorate(I) oxidation as described in Section 10.1.6.

$$CS_2 + 8OCl^- + 2H_2O \longrightarrow CO_2 + 2H_2SO_4 + 8Cl^-$$

For each 0.5 mol (38 g, 30 cm^3) of carbon disulfide, a 25% excess of chlorate(I) is used in the form of 6.7 dm^3 of laundry bleach or a mixture of 550 g of 65% calcium chlorate(I) in 2.2 dm^3 water. The reaction temperature should be regulated at 20-30°C to avoid volatilizing the carbon disulfide.

10.1.8. Carboxylic acids

Incineration or burial in a secure landfill requires only that the acids be collected in plastic or glass containers because of their corrosiveness to metals. Care must be taken to keep them separate from amines and other bases, with which they form salts with the evolution of heat. Water-soluble carboxylic acids and their sodium, potassium, calcium, and magnesium salts can be washed down the drain if local regulations permit.

10.1.9. Other organic acids

Sulfonic acids, RSO_3H, are generally soluble in water and can usually be handled according to the guidelines for carboxylic acids, i.e. incinerated, poured into the drain if local regulations permit, or placed in a secure landfill. Acids of non-hazardous elements, such as phosphonic acids [$RPO(OH)_2$] and boronic acids [$RB(OH)_2$], can be handled like sulfonic acids. Acids of intrinsically hazardous elements, such as arsonic acids [$RAsO(OH)_2$] and stibonic acids [$RSbO(OH)_2$], should be placed in secure landfills.

10.1.10. Acid halides and anhydrides

Acyl halides such as CH_3COCl and C_6H_5COCl, sulfonyl halides such as $C_4H_9SO_2Cl$, and anhydrides such as $(CH_3CO)_2O$ react readily with water, alcohols, and amines. They should never be allowed to come into contact with wastes that contain such substances. Moreover, most of them are sufficiently water-reactive that they cannot be put into lab packs for landfill disposal. Incineration presents no problems other than those associated with halogen-containing compounds.

Most compounds in this class can be hydrolysed to water-soluble products of low toxicity that can be flushed down the drain.

Procedure for hydrolyzing 0.5 mol of $RCOX$, RSO_2X, or $(RCO)_2O$

$$RCOX + 2NaOH \longrightarrow RCO_2Na + NaX + H_2O$$

A 1-dm^3 three-necked flask equipped with a stirrer, dropping funnel, and thermometer is placed on a steam bath in a hood, and 600 cm^3 of 2.5 mol/dm^3 aqueous sodium hydroxide (1.5 mol, 50% excess) are poured into the flask. A few cm^3 of the acid derivative is added dropwise with stirring. If the derivative is a solid, it can be added in small portions through a neck of the flask. If reaction occurs, as indicated by a rise in temperature and solution of the acid derivative, addition is continued without heating of the mixture. If the reaction is sluggish, as may be the case with larger molecules such as p-toluenesulfonyl chloride, the mixture is heated to about 90 °C before adding any more acid derivative. When the initial added material has dissolved, the remainder is added dropwise, and as soon as a clear solution is obtained, the mixture is cooled to room temperature, neutralized to about pH 7 with dilute hydrochloric or sulfuric acid, and washed down the drain with excess water.

10.1.11. Other acid derivatives: esters, amides, nitriles

These compounds can be incinerated or put into a secure landfill. Only a few carboxylic esters are sufficiently toxic to require special handling. β-Propiolactone is an acylating agent that has been shown to be a potent carcinogen by skin-painting tests on animals. However, it is readily hydrolysed to the relatively harmless β-hydroxypropionic acid with aqueous sodium hydroxide (Section 10.1.10), which can be flushed down the drain. Acrylic esters, which may be allergenic, can be hydrolysed to the relatively harmless acrylic acid by ethanolic potassium hydroxide (Section 10.1.2). Methacrylates, which are less toxic, can be treated the same way.

In contrast, sulfonic acid esters and alkyl sulfates are powerful alkylating agents, and some are carcinogenic in animals. They can be readily decomposed by ethanolic potassium hydroxide (Section 10.1.2). Alkyl phosphates and phosphonates can be hydrolysed by the same procedure.

Carboxamides, including the common solvents N,N-dimethylformamide and N,N-dimethylacetamide, are not very hazardous if good laboratory practices are followed. They can be hydrolysed by refluxing for 5 hours with 250 cm^3 of 36% (w/v) hydrochloric acid per mol of amide. Hexamethylphosphoramide (HMPA), a potent carcinogen in animal tests, can be treated the same way, requiring 750 cm^3 of 36% hydrochloric acid per mol of HMPA. The analytical reagent thioacetamide, a moderately active carcinogen in animal tests, is decomposed by 25% excess chlorate (I) as described for thiols (Section 10.1.6).

$$CH_3CSNH_2 + 4OCl^- + H_2O \longrightarrow CH_3CONH_2 + H_2SO_4 + 4Cl^-$$

It is sometimes incorrectly stated that nitriles can be converted into water-soluble and less toxic substances by treating with sodium chlorate(I). Although this treatment is effective for inorganic cyanides, it is not effective for organic nitriles. Nitriles can be converted into carboxylic acids by refluxing with excess ethanolic potassium hydroxide for several hours (Section 10.1.2) or with 250 cm^3 of 36% hydrochloric acid per mol of nitrile for 5-10 hours. Acrylonitrile, a highly toxic compound that should never be poured down the drain, can be hydrolysed by either method.

10.1.12. Aldehydes and ketones

Ketones, especially acetone and methyl ethyl ketone, are very common laboratory solvents. Many aldehydes and ketones are used as intermediates in organic synthesis. All of them burn easily and can be incinerated or burned as fuel supplements.

Many aldehydes are respiratory irritants, and some, such as formaldehyde and acrolein, are quite toxic. Aldehydes may be oxidized to the corresponding carboxylic acids, which are usually less-toxic and less volatile, with aqueous potassium manganate(VII):

$$3RCHO + 2KMnO_4 \longrightarrow 2RCO_2K + RCO_2H + 2MnO_2 + H_2O$$

Procedure for manganate(VII) oxidation of 0.1 mol of aldehyde

A mixture of 100 cm^3 of water and 0.1 mol of aldehyde is stirred in a 1-dm^3 round-bottomed flask equipped with a thermometer, dropping funnel, stirrer, steam bath, and, if the aldehyde boils below 100°C, a condenser. Approximately 30 cm^3 of a solution of 12.6 g (0.08 mol, 20% excess) of potassium manganate(VII) in 250 cm^3 of water are added over a period of 10 minutes. If this addition is not accompanied by a rise in temperature and loss of the purple manganate(VII) colour, the mixture is heated by the steam bath until a temperature is reached at which the colour is discharged. The rest of the manganate(VII) solution is added at within 10 °C of this temperature. The temperature is then raised to 70-80 °C, and stirring continued for an hour or until the purple colour has disappeared, whichever occurs first. The mixture is cooled to room temperature and acidified with 3 mol/dm^3 sulfuric acid. *(CAUTION: Do not add concentrated sulfuric acid to manganate(VII) solution because explosive Mn$_2$O$_7$ may precipitate.)* Enough solid sodium hydrogen sulphate(IV) (at least 8.3 g, 0.08 mol) is added with stirring at 20-40 °C to reduce all the manganese to the oxidation state (II), as indicated by loss of purple colour and dissolution of the solid manganese(IV) oxide. The mixture is washed down the drain with a large volume of water.

If the aldehyde contains a carbon-carbon double bond, as in the case of the highly toxic acrolein, 4 mol (20% excess) of manganate(VII) per mol of aldehyde are used to oxidize the alkene bond and the aldehyde group.

Formaldehyde is conveniently oxidized to formic acid and carbon dioxide by sodium chlorate(I). Thus 10 cm^3 formalin (37% formaldehyde) in 100 cm^3 water is stirred into 250 cm^3 laundry bleach (5.25% NaOCl) at room temperature and allowed to stand for 20 minutes before being flushed down the drain. This procedure is not recommended for other aliphatic aldehydes because it leads to chloro acids, which are more toxic and less biodegradable than corresponding unchlorinated acids.

Most ketones have such low toxicity and are so easily handled that there is no need to convert them into other substances for disposal. However, should degradation be necessary, it may be carried out as follows. Methyl ketones, such as the neurotoxin methyl n-butyl ketone, can be converted into a carboxylic acid with laundry bleach (5.25% NaOCl):

$$RCOCH_3 + 3OCl^- \longrightarrow RCO_2^- + CHCl_3 + 2OH^-$$

An unsaturated methyl ketone, such as methyl vinyl ketone, can be destroyed the same way. In general, most unsaturated ketones can be split at the double bond by the procedure for oxidizing aldehydes given above but using 3.2 mol (20% excess) of manganate(VII) per mol of ketone:

$$3RCOCHCHR + 8KMnO_4 \longrightarrow 3RCOCO_2K + 3RCO_2K + 2KOH + 8MnO_2 + 2H_2O$$

10.1.13. Amines

Aliphatic and aromatic amines are used principally as intermediates in synthesis. A few tertiary amines, such as pyridine and triethylamine, are used as solvents or catalysts. They can be incinerated or put into a secure landfill. They should not be mixed with waste acids.

Aromatic amines are relatively toxic compounds, destroying the oxygen-carrying capacity of haemoglobin. Some, especially those with more than one aromatic ring, have been shown to be carcinogenic to animals and in some cases to man (e.g., 2-naphthylamine, 4-aminobiphenyl). Acidified potassium manganate(VII) efficiently degrades aromatic amines. Nine strongly mutagenic mono- and di-amines have been treated with this reagent, and in no case was amine detectable by chromatography or a mutagenicity test after treatment.

Procedure for manganate(VII) oxidation of 0.01 mol of aromatic amine.

A solution of 0.01 mol of aromatic amine in 3 dm^3 of 1.7 mol/dm^3 sulfuric acid is prepared in a 5-dm^3 flask; 1 dm^3 of 0.2 mol/dm^3 potassium manganate(VII) is added, and the solution allowed to stand at room temperature for 8 hours. Excess manganate(VII) is reduced to the oxidation state(II) by slowly adding solid sodium hydrogen sulfate(IV). The mixture is then flushed down the drain.

10.1.14. Nitro compounds

Both aliphatic and aromatic nitro compounds are relatively toxic, and both can form explosive compounds on exposure to strong bases. Some poly(nitro)arenes are powerful explosives, 2,4,6-trinitrotoluene and picric acid (2,4,6-trinitrophenol) being well known examples. Picric acid, a common laboratory reagent, is considered in Chapter 11.

Most nitro compounds, except those known or suspected to be explosive, should be incinerated or landfilled. Explosives should be disposed of only by trained personnel and detonated under carefully controlled conditions. Some waste-disposal firms specialize in highly reactive materials and will provide this service. Small quantities of picric acid in water can be reduced to non-explosive triaminophenol by tin and hydrochloric acid, a procedure applicable to small amounts of other polynitro compounds.

$$HOC_6H_2(NO_2)_3 \xrightarrow[\text{(ii) NaOH}]{\text{(i) Sn/HCl}} HOC_6H_2(NH_2)_3 \longrightarrow \textbf{Resin in air}$$

Procedure for reduction of one gram of picric acid with tin

One gram of picric acid and 10 cm^3 of water are stirred in a three-necked flask equipped with a condenser, dropping funnel, and stirrer, and immersed in an ice bath. Four grams of granular tin are added, followed dropwise by the addition of 15 cm^3 of 12 mol/dm^3 hydrochloric acid at a rate that will produce a moderate reaction with the tin. The bath is removed and the mixture gradually heated to its boiling point and refluxed for 1 hour. It is then cooled and filtered to separate the tin. The tin is washed with 10 cm^3 of 2 mol/dm^3 hydrochloric acid, and the filtrate, a solution of triaminophenol, is neutralized with 10% sodium hydroxide solution. The slurry of tin hydroxides, which soon turns black from resinification of triaminophenol should be disposed of by landfill.

10.1.15. N-Nitroso compounds

The carcinogenic properties of some of this class of compounds has generated considerable research interest in good methods for destroying small quantities. Incineration is an effective means of destruction. For destruction of milligram quantities or for removing traces from the surface of glassware, a 3% solution of hydrobromic acid in acetic acid at room temperature is effective. For gram quantities, reduction to the amine by aluminium-nickel alloy in alkali is more practical.

Denitrosation of one milligram of a dialkylnitrosamine with HBr

$$R_2NNO + HBr \longrightarrow R_2NH + BrNO$$

A methylene chloride solution containing 1 mg of a dialkylnitrosamine is concentrated to 1-2 cm^3, dried over anhydrous sodium sulfate(VI) (water makes the HBr reagent less effective), and mixed with 5 cm^3 of 3% (w/v) solution of HBr in acetic acid. The latter solution is prepared by diluting commercial 30% (w/v) HBr in acetic acid with glacial acetic acid. The mixture is kept at ambient temperature for 2 hours, and then diluted with water, neutralized with 10% aqueous sodium hydroxide, and flushed down the drain with a large excess of water.

Decomposition of 0.05 mol of a dialkylnitrosamine with aluminium-nickel alloy.

This procedure is applicable to dialkylnitrosamines that are more than 1% soluble in water, which usually means those that contain fewer than 7 carbon atoms. A solution of 0.05 mol of the dialkylnitrosamine in 500 cm^3 of water is prepared in a 2-cm^3 three-necked flask located in a hood and equipped with a stirrer, an ice bath, and an outlet tube leading through plastic tubing to the back of the hood to remove the hydrogen that is evolved. Then 500 cm^3 of 1 mol/dm^3 sodium hydroxide is added, followed by 50 g 50/50 aluminium-nickel alloy powder gradually added in small portions over a period of about an hour. The flask neck through which it is added is stoppered between additions. The reaction is

Laboratory destruction of hazardous chemicals 175

highly exothermic and accompanied by frothing which can be excessive if the alloy is added too rapidly. Stirring is continued for 3 hours in the ice bath and then for 20 hours at ambient temperature. The finely divided black nickel is allowed to settle, and the aqueous phase decanted, neutralized, and flushed down the drain with excess water.

$$R_2NNO \xrightarrow[NaOH]{Al/Ni} R_2NH + NH_3$$

The residual nickel can ignite in air if allowed to dry. Therefore it is washed with 100 cm^3 of water and suspended in 200 cm^3 of water in the three-necked flask, and 800 cm^3 of 1 mol/dm^3 hydrochloric acid are added gradually with stirring. This is continued until the nickel has dissolved and the resulting solution is treated appropriately as described in Section 10.2.2.

This procedure can be used to destroy dialkylnitrosamines that are less than 1% soluble in water by dissolving the 0.05 mol of the dialkylnitrosamine in 500 cm^3 of methanol instead of water.

The aluminium-nickel alloy procedure cannot be used with N-nitrosamides, such as N-nitrosomethylurea and N-methyl-N'-nitrosoguanidine (MNNG), which react with bases to evolve highly toxic and explosive diazoalkanes.

The hydrobromic acid procedure described for N-nitrosodialkylamines can be used for these amides. However, it is usually more convenient to destroy them by treatment with aqueous acid in methanol, which hydrolyses them to harmless products in most cases, but mutagenic residues may be obtained, which imposes limitations on the application of the method.

Procedure for destruction of 15 grams of an nitrosamide

A solution of 15 g of an N-nitrosamide in 500 cm^3 of methanol, acetone, or water is placed in a 2-dm^3 three-necked flask equipped with a stirrer, dropping funnel, and thermometer. While the solution is kept at 20-30 °C by means of an ice bath, 500 cm^3 of 6 mol/dm^3 hydrochloric acid are added with stirring, followed by the gradual addition of 35 g of solid sulfamic acid to destroy the nitric(III) acid released by hydrolysis. The mixture is stirred at room temperature for 24 hours, neutralized with 6 mol/dm^3 sodium hydroxide solution, and flushed down the drain.

10.1.16. Organic peroxides

The structures, properties and storage of these compounds have been discussed in Section 7.2.1. Several common classes of organic compounds can form peroxides on storage (Appendix B), and safe disposal of such unwanted peroxides is a more common problem than that of excess peroxide reagents. The presence of peroxides should be tested for before disposal (Section 7.2.1). If more than traces are detected, then appropriate methods for the disposal of peroxides must be used. It may be convenient to destroy low concentrations in solvents by treatment with alumina, activated molecular sieve pellets, or iron(II) sulfate (Section 7.2.1), after which the solvent can be disposed of by a method suitable for its class.

Peroxide-containing solvents can often be safely incinerated, particularly if the peroxide content is low. Such solvents should not be mixed with other waste materials but kept in their original containers, which are incinerated one at a time. If the solvent contains a high concentration of peroxide, it should first be diluted with the same solvent peroxide-free or with a high boiling solvent such as dimethyl phthalate. The plan for incineration must be worked out in advance with the incinerator operator. Solvents that contain a high concentration of peroxides should be treated before being sent to a secure landfill.

The greatest hazard is presented by a solid material that has crystallized from a peroxidizable solvent or that has been left after evaporation. If the material is in an open vessel it is feasible to add solvent cautiously to dissolve the peroxide. The resultant dilute solution can be treated as described earlier to destroy the peroxide. However, if the material is in a container that has not been opened for some time, such as a screw-capped reagent bottle, there is a possibility that some peroxide may be present in the container and that it could be detonated by the friction of removing the cap. These materials should be handled and destroyed by personnel trained in dealing with explosives, such as a police bomb squad. Several serious accidents have resulted from the explosion of the crystalline peroxide that is formed by diisopropyl ether on exposure to air.

Commonly used peroxide reagents, such as acetyl peroxide, benzoyl peroxide, t-butyl hydroperoxide, and di-t-butyl peroxide, are somewhat less dangerous than the adventitious peroxides formed in solvents, because their composition and properties are known and because they are usually accompanied by information from the manufacturer on safe handling and disposal. These reagents are usually purchased or prepared in small quantities, and the excess can be disposed of by incineration, one container at a time. It is safer if they are first diluted with water or with a high-boiling solvent, such as a phthalate ester. These peroxides can also be burned in small batches in an isolated open ditch if permitted by local regulations. The peroxide is spread thinly on the ground and ignited with a long-handled torch or similar device. Peroxides with self-accelerating decomposition temperatures below room temperature (e.g., diisobutyryl peroxide, diisopropyl peroxydicarbonate) can be spread in the same ditch and allowed to decompose for a day, provided that access to the disposal ditch is carefully controlled. To ensure complete destruction, the residual decomposition products can then be burned. Organic peroxides (except for adventitious traces in solvents) are not permitted in secure landfills.

The sensitivity of most peroxides to shock and heat can be reduced by diluting them with inert solvents, and such dilution is recommended before transportation for disposal. Alkanes are the most commonly used, with phthalate esters or water as typical alternatives if different solvent characteristics are required.

Because of the potential hazard in transporting peroxides and the small quantities involved, they are often most conveniently decomposed in the laboratory. Acidified iron(II) sulfate solution is useful for destroying hydroperoxides, which are added to a solution containing about 50% molar excess iron(II) sulfate with stirring at room temperature:

$$ROOH + 2Fe^{2+} + 2H^+ \longrightarrow ROH + 2Fe^{3+} + H_2O$$

Diacyl peroxides can be destroyed by this reagent as well as by aqueous sodium hydrogen sulphate(IV), sodium hydroxide, or ammonia. However, diacyl peroxides with low solubility in water, such as dibenzoyl peroxide, react very slowly. A better reagent is a solution of sodium iodide or potassium iodide in glacial acetic acid:

$$(RCO_2)_2 + 2NaI \longrightarrow 2RCO_2Na + I_2$$

For 0.01 mol of diacyl peroxide, 0.022 mol (10% excess) of sodium or potassium iodide is dissolved in 70 cm^3 glacial acetic acid, and the peroxide added gradually with stirring at room temperature. The solution is rapidly darkened by the formation of iodine. After 30 minutes the solution is washed down the drain with a large excess of water.

Most dialkyl peroxides (ROOR) do not react readily at room temperature with iron(II) sulfate, iodide, ammonia, or the other reagents mentioned above. However, these peroxides can be destroyed by a modification of the iodide procedure.

One milliliter of 36% (w/v) hydrochloric acid is added to the above acetic acid/potassium iodide solution as an accelerator, followed by dialkyl peroxide (0.01 mol). The solution is heated to 90-100 $^{\circ}$C on a steam bath over the course of a 30 minutes and held at that temperature for 5 hours.

10.1.17 Dyes and pigments

Dyes and organic pigments generally present no problems for either incineration or disposal in a landfill. Even though some are sufficiently water-soluble for drain disposal, this route should be limited to very small quantities in order not to impart colour to the waste stream.

Arenediazonium salts (ArN$_2^+$ X$^-$) are commonly used as intermediates for many kinds of organic synthesis. Some are unstable in solution, and many are explosive in the solid state.

178 *Chemical Safety Matters*

An arenediazonium salt can be disposed of by adding it gradually to a stirred solution of 5-10% excess 2-naphthol in 3% aqueous sodium hydroxide at 0-20 °C. The resulting azo dye is separated by filtration and either incinerated or put into a secure landfill.

10.2. INORGANIC CHEMICALS

The majority of inorganic wastes consist of a cation (metal or metalloid atom) and an anion (which may or may not contain a metalloid component). In planning the disposal of these substances, it is necessary to determine whether they present a sufficiently low hazard to be placed in a sanitary landfill, or the sewer system, or whether the potential hazard is great enough to require disposal in a secure landfill. It is helpful to examine the cationic and anionic parts of the substance separately. If either presents significant potential hazard, the substance should go to a secure landfill (possibly via an incinerator and the ash from there). Some elements can present disposal problems because of reactivity (e.g., sodium) or toxicity (e.g., bromine).

Although it is sometimes assumed that if a substance contains a "heavy metal" it is highly toxic, this is not a sound basis for decision. Whilst salts of some heavy metals, such as lead, thallium, and mercury, are highly toxic, those of others, such as gold and tantalum, are not. On the other hand, compounds of beryllium, a "light metal", are highly toxic. In Table 10.1 cations of metals and metalloids are listed alphabetically in two groups, those whose toxic properties as described in the toxicological literature present a significant hazard, and those whose properties do not. The basis for separation is relative and does not imply that those in the second list are "non-toxic." It is a basic precept of toxicology that every substance, even oxygen or water, can be toxic under certain conditions.

Similarly, Table 10.2 lists anions according to their level of toxicity and other dangerous properties, such as strong oxidizing power [e.g., chlorate(VII)], flammability (e.g., hydride), or explosivity (e.g., azide).

Elements that pose a hazard because of significant radioactivity are outside the scope of this monograph, and none is considered here. Their handling and disposal are highly regulated in most countries.

10.2.1. Chemicals in which neither the cation nor the anion presents significant hazard

These consist of those chemicals composed of ions from the right-hand columns of Tables 10.1 and 10.2. Those that are soluble in water to the extent of a few percent can usually be washed down the drain (Chapter 15). Only laboratory quantities should be disposed of in this manner, and at least 100 parts of water per part of chemical should be used. Local regulations should be checked for possible restrictions on certain

Table 10.1. *Relative toxicities of cations and prefered precipitants.*

High hazard cation	Precipitant[a]	Low hazard cation	Precipitant[a]
Antimony	OH^-, S^{2-}	Aluminium	OH^-
Arsenic	S^{2-}	Bismuth	OH^-, S^{2-}
Barium	SO_4^{2-}, CO_3^{2-}	Calcium	SO_4^{2-}, CO_3^{2-}
Beryllium	OH^-	Cerium	OH^-
Cadmium	OH^-, S^{2-}	Cesium	
Chromium(III)[b]	OH^-	Copper[c]	OH^-, S^{2-}
Cobalt(II)[b]	OH^-, S^{2-}	Gold*	OH^-, S^{2-}
Gallium	OH^-	Iron[c]	OH^-, S^{2-}
Germanium	OH^-, S^{2-}	Lanthanides	OH^-
Hafnium	OH^-	Lithium	
Indium	OH^-, S^{2-}	Magnesium	OH^-
Iridium*	OH^-, S^{2-}	Molybdenum (VI)[b,d]	
Lead	OH^-, S^{2-}	Niobium(V)	OH^-
Manganese(II)[b]	OH^-, S^{2-}	Palladium*	OH^-, S^{2-}
Mercury	OH^-, S^{2-}	Potassium	
Nickel	OH^-, S^{2-}	Rubidium	
Osmium(IV)*[b,e]	OH^-, S^{2-}	Scandium	OH^-
Platinum(II)*[b]	OH^-, S^{2-}	Sodium	
Rhenium(VII)*[b]	S^{2-}	Strontium	SO_4^{2-}, CO_3^{2-}
Rhodium(III)*[b]	OH^-, S^{2-}	Tantalum	OH^-
Rutheniwm(III)*[b]	OH^-, S^{2-}	Tin	OH^-, S^{2-}
Selenium	S^{2-}	Titanium	OH^-
Silver*	Cl^-, OH^-, S^{2-}	Yttrium	OH^-
Tellurium	S^{2-}	Zinc[c]	OH^-, S^{2-}
Thallium	OH^-, S^{2-}	Zirconium	OH^-
Tungsten(VI)[b,d]			
Vanadium	OH^-, S^{2-}		

[a] Precipitants are listed in order of preference:
OH^- = base (sodium hydroxide or sodium carbonate), S^{2-} = sulfide, and Cl^- = chloride
[b] The precipitant is for the indicated valence state.
[c] Maximum tolerance levels have been set for these low-toxicity ions in some countries, and large amounts should not be put into public sewer systems. The small amounts typically used in laboratories will not normally affect water supplies.
[d] These ions are best precipitated as calcium molybdate(VI) or calcium tungstate(VI).
[e] CAUTION: Osmium tetroxide, OsO_4, a volatile, extremely poisonous substance, is formed from almost any osmium compound under acid conditions in the presence of air.
* Recovery of these rare and expensive metals may pay for itself.

180 Chemical Safety Matters

Table 10.2. Relative toxicity of anions and prefered precipitants

High hazard anions	Ion type[a]	Precipitant	Low-hazard anions
Aluminium hydride, AlH_4^-	F	--	Hydrgen sulfate(IV), HSO_3^-
Amide, NH_2^-	F,E[b]	--	Borate, BO_3^{3-}, $B_4O_7^{2-}$
Arsenate(V), AsO_3^-, AsO_4^{3-}	T	Cu^{2+}, Fe^{2+}	Bromide, Br^-
Arsenate(III), AsO_2^-, AsO_3^{3-}	T	Pb^{2+}	Carbonate, CO_3^{2-}
Azide, N_3^-	E, T	-	Chloride, Cl^-
Borohydride, BH_4^-	F	-	Cyanate, OCN^-
Bromate(V), BrO_3^-	O, F, E	-	Hydroxide, OH^-
Chlorate(I), OCl^-	O	-	
Chlorate(V), ClO_3^-	O, E	-	Iodide, I^-
Chlorate(VII), ClO_4^-	O, E	-	
Chromate(VI), CrO_4^{2-}, $Cr_2O_7^{2-}$	T, O[c]	-	Oxide, O_2^-
Cyanide, CN^-	T	-	Phosphate, PO_4^{3-}
Ferricyanide, $[Fe(CN)_6]^{3-}$	T	Fe^{2+}	Sulfate(VI), SO_4^{2-}
Ferrocyanide, $Fe(CN)_6]^{4-}$	T	Fe^{3+}	Suflate(IV), SO_3^{2-}
Fluoride, F^-	T	Ca^{2+}	Thiocyanate, SCN^-
Hydride, H^-	F	-	
Hydroperoxide, O_2H^-	O, E	-	
Hydrogen sulfide, SH^-	T	-	
Iodate(V), IO_3^-	O, E	-	
Manganate(VII), MnO_4^-	T, O[d]		
Nitrate(III), NO_2^-	T, O		
Nitrate(V), NO_3^-	O		
Peroxide, O_2^{2-}	O, E	-	
Peroxodisulfate(VI), $S_2O_8^{2-}$	I	-	
Selenate, SeO_4^{2-}	T	Pb^{2+}	
Selenide, Se^{2-}	T	Cu^{2+}	
Sulfide, S^{2-}	T[e]		

[a] Toxic, T; oxidant, O; flammable, F; explosive, E.
[b] Metal amides readily form explosive peroxides on exposure to air (Section 10.2.3).
[c] Reduce and precipitate as Cr(III); see Table 10.1.
[d] Reduce and precipitate as Mn(II); see Table 10.1.
[e] See Table 10.3.

elements, such as copper. Dilute slurries of insoluble materials, such as calcium sulfate or aluminium oxide, also can be handled in this way, provided the material is finely divided and not contaminated with tar which might clog the piping. Some incinerators can handle these chemicals.

When these options are not available, disposal must be in a landfill. For chemicals of this type, a sanitary landfill can be used if local regulations permit. However, if the solution or suspension is dilute, the cost of landfilling may be uneconomic. If time and space permit, dilute aqueous solutions can be boiled down or allowed to evaporate to leave only a sludge of the inorganic solid for landfill disposal.

An alternative procedure is to precipitate the metal ion by the agent recommended in Table 10.1 and send the precipitate to a landfill. The most generally applicable procedure is to precipitate the cation as the hydroxide or oxide by adjusting the pH to the range shown in Appendix F. Because the pH range for precipitation varies greatly among metal ions, it is important to control pH carefully. The aqueous solution of the metal ion is adjusted to the recommended pH by addition of 1 mol/dm^3 sulfuric acid, or sodium hydroxide or carbonate. The pH can be determined over the range 1-10 by use of wide-range pH paper. For some ions, the hydroxide precipitate will redissolve at a high pH. For a number of metal ions the use of sodium carbonate will result in precipitation of the metal carbonate or a mixture of hydroxide and carbonate.

The precipitate is separated by filtration, or as a heavy sludge by decantation, and packed for landfill disposal. Some gelatinous hydroxides are difficult to filter. In such cases, heating the mixture close to 100 °C or stirring with diatomaceous earth, approximately 1-2 times the weight of the precipitate, often facilitates filtration. As shown in Table 10.1, precipitants other than a base may be superior for some metal ions, e.g., sulfuric acid for calcium ion.

10.2.2. Chemicals in which a cation presents a relatively high hazard from toxicity

In general, waste chemicals containing any of the cations listed under as high hazardous in Table 10.1 should be disposed of in a secure landfill. In many cases the waste can be packaged in a lab pack without treatment. Economics dictate whether an aqueous solution of a salt of a toxic cation is converted into an insoluble derivative in order to reduce the volume of material in the lab pack.

The methods of concentrating or separating metal ions described in Section 10.2.1 also apply here. Most of the high toxic hazard cations listed in Table 10.1 can be separated from aqueous solutions as the hydroxide or oxide (Appendix F). Alternatively, many can be precipitated as insoluble sulfides by treatment with sodium sulfide in neutral solution. Several sulfides will redissolve in excess sulfide ion, so it is important that the sulfide ion concentration be controlled by adjustment of the pH.

Precipitation is achieved at a neutral pH by the addition of 1 mol/dm^3 sodium hydroxide or 1 mol/dm^3 sulfuric acid, using wide-range pH paper. A 1 mol/dm^3 solution of sodium sulfide is added to the metal ion solution, and the pH is again adjusted to neutral with sulfuric acid. The precipitate is separated by filtration or decantation and packed for landfill disposal. Excess sulfide ion can be destroyed by the addition of chlorate(I) to the

Table 10.3. Precipitation of sulfides

Precipitated at pH 7	Not precipitated at low pH	Soluble complex at high pH
Ag(I)		
As(III)[a]		X
Au(I)[a]		X
Bi(III)		
Cd(II)		
Co(II)	X	
Cr(II)[a]		
Cu(II)		
Fe(II)[a]	X	
Ge(II)		X
Hg(II)		X
In(III)	X	
Ir(IV)		X
Mn(II)[a]	X	
Mo(II)		X
Ni(II)	X	
Os(IV)		
Pb(II)		
Pd(II)[a]		
Pt(II)[a]		X
Re(IV)		
Rh(II)[a]		
Ru(IV)		
Sb(III)[a]		X
Se(II)		X
Sn(II)		X
Te(IV)		X
Tl(I)[a]	X	
V(IV)[a]		
Zn(II)	X	

[a] Higher oxidation states of this ion are reduced by sulfide ion and precipitated as this sulfide.

clear aqueous phase (Section 10.2.3). Guidance on the precipitation of many cations as sulfides is provided in Table 10.3.

The following ions are most commonly found as oxyanions and are not precipitated by base: As(III), As(V), Re(VII), Se(IV), Se(VI), Te(IV) and Te(VI). These elements can be precipitated from their oxyanions as the sulfides by the above procedure. Oxyanions of Mo(VI) and W(VI) can be precipitated as their calcium salts by the addition of calcium chloride. Some toxic ions can be absorbed by passing their solutions over ion-exchange resins. The resins can be landfilled, and the effluent solutions poured down the drain.

Another class of compounds whose cations may not be precipitated by the addition of hydroxide ions are the most stable complexes of metal cations with Lewis bases such as ammonia, amines, or tertiary phosphines. Because of the large number of these compounds and their wide range of properties, it is not possible to give a general procedure for separating the cations. In many cases metal sulfides can be precipitated directly from aqueous solutions of the complexes by the addition of aqueous sodium sulfide. If a test-tube experiment shows that stronger measures are needed, the addition of hydrochloric acid to produce a slightly acidic solution will often decompose the complex by protonation of the basic ligand. Metal ions that form insoluble sulfides under acid conditions can then be precipitated by dropwise addition of aqueous sodium sulfide.

A third option for these wastes is incineration, provided that the incinerator ash is to be sent to a secure landfill. Incineration to ash reduces the volume of wastes going to a landfill. Wastes that contain mercury, thallium, gallium, osmium, selenium or arsenic should not be incinerated because volatile, toxic combustion products may be emitted.

10.2.3. Chemicals in which an anion presents a relatively high hazard

The more common dangerous anions are listed in Table 10.2. Many of the comments made previously about the disposal of dangerous cations apply to these anions. The hazard associated with some of these anions is their reactivity or potential to explode, which makes them unsuitable for landfill disposal. Others can be placed in lab packs for landfill disposal. Most chemicals containing these anions can be incinerated, but strong oxidizing agents and hydrides should be introduced into the incinerator only in containers of not more than a few hundred grams. Incinerator ash from anions of chromium or manganese should be transferred to a secure landfill.

Some of these anions can be precipitated as insoluble salts for landfill disposal, as indicated in Table 10.2. Small amounts of strong oxidizing agents, hydrides, cyanides, azides, metal amides, and soluble sulfides or fluorides can be converted into less hazardous substances in the laboratory before being disposed of. Suggested procedures are presented in the following paragraphs.

Oxidizing agents

Chlorates(I), chlorates(V), bromates(V), iodates(V), iodates(VII), inorganic peroxides and hydroperoxides, sulfates(VII), chromates(VI), molybdates(VI) and manganates(VII) can be reduced by sodium hydrogen sulfate(IV).

A dilute solution or suspension of a salt containing one of these anions has its pH reduced to less than 3 with sulfuric acid, and a 50% excess of aqueous sodium hydrogen sulfate(IV) is added gradually with stirring at room temperature. An increase in temperature indicates that the reaction is taking place. If the reaction does not start on addition of about 10% of the sodium hydrogen sulfate(IV), a further reduction in pH may initiate it. Coloured anions (e.g., manganate(VII), chromate(VI)) serve as their own indicators of completion of the reduction. The reduced mixtures can usually be washed down the sink. However, if large amounts of manganate(VII) or chromate(VI) have been reduced, it may be necessary to transfer to a secure landfill, possibly after a reduction in volume by concentration or precipitation as described above in Section 10.2.1.

Hydrogen peroxide can be reduced by the sodium hydrogen sulfate(IV) procedure or by iron(II) sulfate as described for organic hydroperoxides (Section 10.1.16). However, it is usually acceptable to dilute it to a concentration less than 3% and flush down the drain. Solutions with a hydrogen peroxide concentration greater than 30% should be handled with great care to avoid contact with reducing agents, including all organic materials, or with transition metal compounds, which can catalyse violent reactions.

Concentrated chloric(VII) acid (particularly when stronger than 60%) must be kept away from reducing agents, including weak reducing agents such as ammonia, wood, paper, plastics, and all other organic substances, because it can react violently with them. Dilute chloric(VII) acid is not reduced by common laboratory reducing agents such as sodium hydrogen sulfate(IV), hydrogen sulfide, hydriodic acid, iron, or zinc. Chloric(VII) acid is most easily disposed of by stirring it gradually into enough cold water to make its concentration less than 5%, neutralizing it with aqueous sodium hydroxide, and washing the solution down the drain with a large excess of water.

Nitrate(V) is most dangerous in the form of concentrated nitric(V) acid (70% or higher), which is a potent oxidizing agent for organic materials and all other reducing agents. It can also cause serious skin burns. Dilute aqueous nitric(V) acid is not a dangerous oxidizing agent and is not easily reduced by common laboratory reducing agents. Nitric(V) acid should be neutralized with aqueous sodium hydroxide before disposal down the drain; concentrated nitric(V) acid should be diluted before neutralization. Metal nitrates(V) are generally quite soluble in water. Those of the metals listed under low toxic hazard in Table 10.1, as well as ammonium nitrate(V), should be kept separate from oil or other organic materials because on heating such a combination fire or explosion can occur. Otherwise, these can be treated as chemicals that present no significant hazard.

Nitrates(III) in aqueous solution can be destroyed by adding about 50% excess aqueous ammonia and acidifying with hydrochloric acid to pH 1.

$$HNO_2 + NH_3 + H^+ \longrightarrow N_2 + 2H_2O$$

Metal hydrides

Most metal hydrides react violently with water with the evolution of hydrogen, which can form an explosive mixture with air. Some, such as sodium hydride and lithium aluminium hydride are pyrophoric. Most can be decomposed by gradual addition of methanol, ethanol, n-butyl alcohol, or t-butyl alcohol to a stirred, ice-cooled solution or suspension of the hydride in an inert liquid, such as diethyl ether, tetrahydrofuran, or toluene, under nitrogen in a three-necked flask.

Four common hydrides in laboratories are lithium aluminium hydride, sodium borohydride, sodium hydride, and calcium hydride. The following methods for their disposal demonstrate that the reactivity of metal hydrides varies considerably. Most hydrides can be safely decomposed by one of the four methods, but the properties of a given hydride must be well understood in order to select the most appropriate method.

Lithium aluminium hydride (LiAlH$_4$) can be purchased as a solid or as a 0.5-1.0 mol/dm^3 solution in toluene, diethyl ether, tetrahydrofuran, or other ethers. Although dropwise addition of water to its solutions under nitrogen in a three-necked flask has frequently been used to decompose it, vigorous frothing often occurs. An alternative is to use 95% ethanol, which reacts less vigorously than water. A safer procedure is to decompose the hydride with ethyl acetate, because no hydrogen is formed:

$$2CH_3CO_2C_2H_5 + LiAlH_4 \longrightarrow LiOC_2H_5 + Al(OC_2H_5)_3$$

The mixture sometimes becomes so viscous after the addition of ethyl acetate that stirring is difficult. When the reaction with ethyl acetate has ceased, a saturated aqueous solution of ammonium chloride is added with stirring. The mixture separates into an organic layer and an aqueous layer containing inert inorganic solids. The upper, organic layer should be packed for incineration or landfill disposal. The lower, aqueous layer can be flushed down the drain.

Sodium borohydride (NaBH$_4$) is so stable in water that a 12% aqueous solution stablized with sodium hydroxide is sold commercially. In order to effect decomposition, the solid or aqueous solution is added to enough water to make the borohydride concentration less than 3%, and then excess dilute aqueous acetic acid is added dropwise with stirring under nitrogen.

Sodium hydride (NaH) in the dry state is pyrophoric, but it can be purchased as a relatively safe dispersion in mineral oil. Either form can be decomposed by adding enough dry hydrocarbon solvent (e.g., heptane) to reduce the hydride concentration below 5% and then adding excess t-butyl alcohol dropwise under nitrogen with stirring. Cold water is then added gradually, and the two layers separated. The organic layer can be incinerated or sent to a secure landfill, and the aqueous layer can be flushed down the drain.

Calcium hydride (CaH_2) is purchased as a powder, usually for drying ethers, esters, alcohols with more than three carbon atoms, and other solvents. It is decomposed by adding 25 cm^3 of methanol per gram of hydride under nitrogen with stirring. When the reaction has finished, an equal volume of water is gradually added to the stirred slurry of calcium methoxide, and the mixture is then flushed down the drain with a large volume of water.

Inorganic sulfides

Small amounts of unused sodium sulfide or potassium sulfide can be destroyed in aqueous solution by sodium or calcium chlorate(I), using the procedure described for oxidizing thiols (Section 10.1.6).

$$Na_2S + 4OCl^- \longrightarrow Na_2SO_4 + 4Cl^-$$

Small amounts of hydrogen sulfide can be oxidized by the same reagents. To keep the reaction under control, the hydrogen sulfide is absorbed in excess aqueous sodium hydroxide, and then chlorate(I) solution is added dropwise in a hood. (CAUTION: Hydrogen sulfide is an acute poison. Section 8.4.9).

An alternative method is to precipitate the sulfide ion as insoluble iron(II) sulfide (Table 10.3) for landfill disposal.

Inorganic fluorides

Each worker should be familiar with the properties and proper handling of hydrogen fluoride (HF) before working with it (Section 6.8.2). Hydrogen fluoride can cause serious burns, with the hazard from aqueous solutions increasing with concentration. Waste aqueous hydrofluoric acid can be added slowly to a stirred solution of excess slaked lime to precipitate calcium fluoride, which is chemically inert and non-toxic. If no more than about 100 g of calcium fluoride are formed, the suspension can be washed down the drain. Larger amounts should be separated and sent to a landfill.

Alternatively, hydrofluoric acid can be diluted to 1% concentration by adding it to cold water in a polyethylene or borosilicate glass vessel, neutralizing the acid with aqueous sodium hydroxide, and precipitating calcium fluoride by addition of excess calcium chloride solution.

Soluble metal fluorides such as sodium fluoride and potassium fluoride are highly poisonous. They can be converted into calcium fluoride by treating aqueous solutions with calcium chloride solution. Boron(III) fluoride can be dissolved in water and the fluoride ion precipitated as calcium fluoride by adding calcium chloride solution. The calcium fluoride is filtered and sent to a landfill, and the filtrate is flushed down the drain.

$$2BF_3 + 3CaCl_2 + 6H_2O \longrightarrow 3CaF_2 + 2B(OH)_3 + 6HCl$$

Inorganic cyanides

Hydrogen cyanide, b.p. 26 °C, is among the most poisonous chemicals known and is toxic by inhalation, ingestion, or skin absorption. Precautions for its safe handling are described in Section 6.8.3. It can be disposed of by dilution with one or more volumes of ethanol followed by incineration of the solution. Small amounts can be oxidized to the relatively harmless cyanate by aqueous sodium chlorate(I), using the procedure described for oxidizing thiols (Section 10.1.6). The oxidation should be carried out carefully in a good hood.

$$NaCN + NaOCl \longrightarrow NaOCN + NaCl$$

Hydrogen cyanide is dissolved in several volumes of ice water in an ice-cooled, three-necked flask equipped with a stirrer, thermometer, and dropping funnel. *(CAUTION: Sodium hydroxide or other bases, including sodium cyanide, must not be allowed to come into contact with liquid hydrogen cyanide because they may initiate a violent polymerization of hydrogen cyanide.)* Approximately 1 molar equivalent of aqueous sodium hydroxide is added at 4-10 °C to convert the hydrogen cyanide into its sodium salt. A 50% excess of commercial laundry bleach [e.g., Clorox®, containing 5.25% (0.75 mol/dm^3 molar) sodium chlorate(I)] is added at 4-10 °C. When the addition is complete and heat is no longer being evolved, the solution is allowed to warm to room temperature and stand for several hours. The mixture is washed down the drain with excess water.

The same procedure can be applied to soluble cyanides, such as sodium cyanide and potassium cyanide, as well as to insoluble cyanides such as copper(I) cyanide. The procedure also destroys soluble ferrocyanides and ferricyanides. Alternatively, these can be precipitated as the iron(III) or iron(II) salt, respectively, for landfill disposal. In calculating the quantity of chlorate(I) required, remember that additional equivalents may be needed if the metal ion can be oxidized to a higher valence state, as in the reaction:

$$2CuCN + 3OCl^- + H_2O \longrightarrow 2Cu^{2+} + 2OCN^- + 2OH^- + 3Cl^-$$

To destroy cyanogen bromide, a 0.6 mol/dm^3 aqueous solution is treated first with an equal volume of 1 mol/dm^3 sodium hydroxide and then with two volumes of laundry bleach.

Metal azides

Heavy metal azides are notoriously explosive. Sodium azide, a common preservative in clinical laboratories and a useful reagent in synthetic work, is only explosive when heated near its decomposition temperature (300 °C). The heating of sodium azide should be avoided. Sodium azide should never be flushed down the drain. This practice has caused serious accidents because the azide can react with lead or copper in the drain lines to produce azide that later explodes. Moreover, sodium azide has high acute toxicity to mammals as well as bacteria in water-treatment plants. It can be destroyed by reaction with nitric(III) acid:

$$2NaN_3 + 2HNO_2 \longrightarrow 3N_2 + 2NO + 2NaOH$$

The operation must be carried out in a hood because of the formation of nitric(II) oxide. An aqueous solution containing no more than 5% sodium azide is put into a three-necked flask equipped with a stirrer, a dropping funnel, and an outlet with plastic tubing to carry nitrogen and nitrogen oxide to the hood flue. [*CAUTION: The order of addition is essential. If the acid is added before the nitrite(III), poisonous, volatile hydrazoic acid (HN_3) will be evolved.*] A 20% aqueous solution of sodium nitrate(III) containing 1.5 g (about 40% excess) of sodium nitrate(III) per gram of sodium azide is added with stirring. A 20% aqueous solution of sulfuric acid is then added gradually until the reaction mixture is acidic to litmus paper. When the evolution of nitrogen oxides ceases, the acidic solution is tested with starch iodide paper. If it turns blue, excess nitrate(III) is present and the decomposition is complete. The reaction mixture is washed down the drain.

Methods for the decomposition of lead azide and the decontamination of drain lines suspected of containing lead azide or copper azide have been described, using nitric(III) acid or cerium(IV) ammonium nitrate(V). However, such potentially dangerous operations should be carried out only by personnel with experience in handling explosives and never by laboratory workers.

Other alkali metal azides can be treated like sodium azide.

Hydrazoic acid (HN_3), besides being highly toxic, is highly explosive and should never be isolated. Solutions of hydrazoic acid in water, benzene, or chloroform are quite stable. Solutions of it in benzene or chloroform can be decomposed by adding enough water to make the concentration of hydrazoic acid in the water less than 5% and stirring vigorously or shaking in a separating funnel to extract the HN_3 into the aqueous layer. The aqueous solution is neutralized with aqueous sodium hydroxide, separated from the organic layer, and treated by the procedure for decomposing sodium azide.

Metal amides

It is difficult to give general directions for disposing of this large, diverse group; most of them are encountered only rarely. Only personnel experienced in handling such substances should work with explosive amides. Those that are not inherently explosive can be incinerated.

Sodium amide and potassium amide are common reagents that are relatively safe to use and dispose of if proper precautions are taken. They may ignite or explode on heating or grinding in air because of explosive

Laboratory destruction of hazardous chemicals

oxidation products that may form on the surface, particularly if they have been previously exposed to air or moisture. The presence of such oxidation products is indicated by a tan, brown, or orange colour, contrasting with the usual white or light grey colour of the unoxidized amide. Any such discoloured amide should be decomposed by the hydrocarbon-ethanol procedure given below. The risk of explosion can be avoided by preparing these amides in the vessel in which they will be used just before they are needed. Sodium amide, free of explosive peroxides, can be purchased in sealed bottles. The bottles should be opened only once, to take out the quantity of reagent needed, and the remainder should be disposed of promptly. It has been recommended that sodium amide be decomposed by covering it with toluene or kerosene and adding 95% ethanol slowly with stirring. Alternatively, sodium amide or potassium amide in small portions can be stirred into excess solid ammonium chloride:

$$NaNH_2 + NH_4Cl \longrightarrow NaCl + 2NH_3$$

10.2.4. Metals

Alkali metals

Alkali metals react violently with water, with common hydroxylic solvents, and with halogenated hydrocarbons. Potassium can form explosive peroxides on exposure to air. Because of these properties, alkali metals cannot be put into landfills. Bottles containing a few grams can be incinerated.

Waste sodium is readily destroyed with 95% ethanol. The procedure is carried out in a three-necked, round-bottomed flask equipped with a stirrer, dropping funnel, water-cooled condenser, and heating mantle or steam bath. Solid sodium should be cut into small pieces while wet with a hydrocarbon, preferably mineral oil, so that the unoxidized surface is exposed. A dispersion of sodium in mineral oil can be treated directly. The flask is flushed with nitrogen and the sodium placed in it. Then 13 cm^3 of 95% ethanol per gram of sodium are added at a rate that causes rapid refluxing. *(CAUTION: Hydrogen is evolved.)* Stirring is commenced as soon as enough ethanol has been added to make this possible. The mixture is stirred and heated under reflux until the sodium is dissolved. The heat source is removed, and an equal volume of water added at a rate that causes no more than mild refluxing. The solution is then cooled, neutralized with 6 mol/dm^3 sulfuric or hydrochloric acid, and washed down the drain.

Lithium metal can be treated by the same procedure, but using 30 cm^3 of 95% ethanol per gram of lithium. The rate of solution is lower than that of sodium.

To destroy metallic potassium, the less reactive t-butyl alcohol is used in the proportion of 21 cm^3 per gram of metal. *(CAUTION: Potassium metal that has formed a yellow oxide coating from exposure to air should not be cut with a knife, even when wet with a hydrocarbon, because of the explosivity of the peroxide.*) If the potassium is dissolving too slowly, a few per cent of methanol can be added gradually to the refluxing t-butyl alcohol. Oxide-coated potassium sticks should be put directly into the flask and decomposed with t-butyl alcohol. The decomposition will require considerable time because of the low surface/volume ratio of the metal sticks.

Other metals

Most metals are pyrophoric if sufficiently finely divided. Above a critical ratio of surface area to mass that is specific for each metal, the heat of oxide formation on exposure to air can lead to ignition or to a dust explosion. This property is well recognized for aluminium, cobalt, iron, magnesium, manganese, nickel, palladium, platinum, titanium, tin, uranium, zinc, zirconium, and their alloys. Any of these metals in finely divided form should be kept under an inert atmosphere in a tightly closed container, and should be exposed to air as little as possible. A fire extinguisher that has a special granular formulation designed to control burning metal should be close at hand.

These metal powders can be disposed of by adding sufficient water to make a paste, which is then spread, a few millimeters thick, on a metal pan to dry in the open air. The pan should be located in a draught-free area and where no harm will result if the powder should ignite. As the paste dries, the surface of the metal particles gradually oxidizes. The resulting metal oxide is no longer pyrophoric and can be recovered or transferred to a container for disposal in a landfill.

Nickel, palladium, platinum, and other metals or combinations that are used as hydrogenation catalysts should never be allowed to dry in air following a hydrogenation (for example on filter paper), because they are prone to ignite. Filter cakes containing such catalysts should be promptly moistened with water. Used nickel catalysts can be dissolved in hydrochloric acid as outlined in Section 10.1.15. Other metal catalysts require special precedures as indicated in Section 9.1.3.

Water-reactive metal halides

These compounds should not be placed in a landfill because of their hazardous reactivity with water (Appendix A). Liquid halides, such as $TiCl_4$ and $SnCl_4$, can be added to well stirred water in a round-bottomed flask cooled by an ice bath as necessary to keep the exothermic reaction under control. It is usually more convenient to add solid halides, such as $AlCl_3$ and $ZrCl_4$, to stirred water and crushed ice in a beaker. The metal ions can be separated and disposed of as described in Section 10.2.1.

10.2.5. Halides and acid halides of non-metals

Halides and acid halides such as BCl_3, PCl_3, $SiCl_4$, $SOCl_2$, SO_2Cl_2, and $POCl_3$ are mostly water-reactive and must not be placed in landfills (Appendix A). The liquids can be conveniently hydrolysed with 2.5 mol/dm^3 sodium hydroxide by the procedure in Section 10.1.10, with the hydrolysate flushed down the drain after neutralization. These compounds are irritating to the skin and respiratory passages and, even more than most chemicals, require a good hood and skin protection when handling them. Moreover, PCl_3 may give off small amounts of highly toxic phosphine (PH_3) during hydrolysis.

Sulfur(I) chloride (S_2Cl_2) is a special case. It is hydrolysed to a mixture of sodium sulfide and sodium sulfate(IV), so that the hydrolysate must be treated with chlorate(I) as described for metal sulfides (Section 10.2.3) before it can be flushed down the drain.

The solids of this class (e.g., PCl_5) tend to cake and fume in moist air, and therefore are not conveniently hydrolysed in a three-necked flask. It is preferable to add them to a 50% excess of 2.5 mol/dm^3 sodium hydroxide solution in a beaker equipped with a stirrer and half-filled with crushed ice. If the solid has not all dissolved by the time the ice has melted and the stirred mixture has reached room temperature, the reaction can be completed by heating on a steam bath.

Boron(III) fluoride can be converted into calcium fluoride (Section 10.2.3).

10.2.6. Non-metal hydrides

Non-metal hydrides, such as B_2H_6, PH_3, and AsH_3, are sensitive to oxidation by many oxidizing agents, and many are pyrophoric and toxic (Appendix C). They can be oxidized to safer materials by aqueous copper(II) sulfate as described for white phosphorus (Section 10.2.7). To avoid violent reaction with air, the oxidation must be carried out under nitrogen, most conveniently in a three-necked flask equipped with stirrer, nitrogen inlet, and dropping funnel or, if the hydride is gaseous, a gas inlet. The reaction mixture is worked up as described for white phosphorus.

10.2.7. Phosphorus

White phosphorus is usually sold in the form of pellets weighing a few grams and stored under water to prevent oxidation. It reacts with most oxidizing agents, including air, in which it is spontaneously flammable (Appendix C). If local regulations permit, gram quantities can be disposed of by being put in a pit in an open field and allowed to burn. Larger quantities can be incinerated by introducing small bottles of phosphorus covered with water into an incinerator one at a time.

In the laboratory, gram quantities of phosphorus can be oxidized by 1 mol/dm^3 aqueous copper(II) sulfate, using twice the required stoichiometric quantity:

$$2P + 5Cu^{2+} + 8H_2O \longrightarrow 2H_3PO_4 + 5Cu + 10H^+$$

The actual reaction is more complex than shown here. The phosphorus becomes coated with black copper phosphide that gradually forms crystalline copper and soluble phosphoric(III) and phosphoric(V) acids.

Procedure for oxidizing five grams of white phosphorus

Five grams (0.16 mol) of white phosphorus are cut under water into pellets up to 5 mm across. The pellets are added to 800 cm^3 (0.80 mol) of 1 mol/dm^3 copper(II) sulfate solution in a 2-dm^3 beaker in a hood. The mixture is allowed to stand for about a week with occasional stirring. The phosphorus gradually disappears, and a fine black precipitate of copper and copper phosphide is formed. The reaction is complete when no waxy white phosphorus is observed on cutting of the larger black pellets under water. The precipitate is separated in a Buchner funnel in a hood and, while still wet, transferred to 500 cm^3 of laundry bleach solution (5.25% NaOCl) and stirred for an hour to ensure the oxidation of any copper phosphide, a potential source of toxic phosphine gas, to phosphate. The copper salt solution is treated as outlined in Section 10.2.1.

Red phosphorus is not pyrophoric and is much safer to work with than white phosphorus. Nevertheless, it is flammable and can create an explosion if mixed with a strong oxidant such as potassium manganate(VII). Red phosphorus can be disposed of by oxidation to phosphoric(V) acid by aqueous potassium chlorate(V).

Procedure for oxidizing five grams of red phosphorus

$$6P + 5KClO_3 + 9H_2O \longrightarrow 6H_3PO_4 + 5KCl$$

Five grams (0.16 mol) of red phosphorus are added to a solution of 33 g (0.27 mol, 100% excess) of potassium chlorate(V) in 2 dm^3 of 1 mol/dm^3 sulfuric acid. The mixture is heated under reflux until the phosphorus has dissolved (usually 5-10 hours, depending on the particle size of the phosphorus). The solution is cooled to room temperature, treated with about 14 g of sodium hydrogen sulfate(IV) to reduce the excess chlorate (Section 10.2.3), and washed down the drain.

10.2.8. Phosphorus(V) oxide

This water-reactive substance can be disposed of by gradually adding it to a mixture of water and crushed ice in a beaker equipped with a stirrer. If there is still unreacted oxide present when the mixture has reached room temperature, it should be stirred and heated on a steam bath until all is dissolved. The solution is neutralized, diluted, and flushed down the drain with excess water.

10.2.9. Hydrazine and substituted hydrazines

Hydrazine is a common reagent that is a carcinogen in animal tests and a strong skin irritant. It can be destroyed by a 25% excess of chlorate(I), allowing a reaction time of 12 hours following the procedure used for oxidizing thiols (Section 10.1.6). The hydrazine should be diluted to a concentration of about 5% with water before commencing the reaction to ensure that it will not become too vigorous.

$$N_2H_2 + 2OCl^- \longrightarrow N_2 + 2Cl^- + 2H_2O$$

Monosubstituted hydrazines, such as phenylhydrazine and methylhydrazine, can be decomposed in similar fashion.

$$RNHNH_2 + OCl^- \longrightarrow RH + N_2 + Cl^- + H_2O$$

Aluminium nickel alloy, used as in the decomposition of N-nitrosamines (Section 10.1.15), is a good alternative for decomposing hydrazine and substituted hydrazines. This alloy is the preferred reagent for destroying 1,1-dimethylhydrazine, for although chlorate(I) destroys most of it, a small yield of carcinogenic N,N-dimethylnitrosamine may appear in the reaction product.

10.3. ORGANO-INORGANIC CHEMICALS

This class includes organometallics, i.e., compounds with metal-carbon bonds such as alkyllithiums, alkylaluminiums, dicyclopentadienyliron (ferrocene), and metal carbonyls. Many of these are sensitive to air and/or water. It also includes compounds with metal atoms bound to organic molecules through oxygen, nitrogen, or other non-metals. Most of these are fairly insensitive to air and water, for example metal alkanecarboxylates, metal alkoxides, metal acetylacetonates, and metal phthalocyanines.

Organo-inorganic chemicals that are insensitive to air or water can be disposed of by incineration or landfilling. If the metal involved is relatively toxic (Table 10.1), the compound or its ash must be transferred to a secure landfill. Appendices A and C list the more common kinds of organo-inorganic compounds that are water-reactive or pyrophoric, respectively. Information on their disposal in the laboratory is given below.

10.3.1. Organometallics sensitive to air or water

Common examples of this group are Grignard reagents, such as alkyllithiums or aryllithiums, and trialkylaluminiums, which are prepared, purchased, or used as dilute solutions in an ether or hydrocarbon. Less-common examples are arylsodiums and dialkylzincs. By contrast the trialkylaluminiums and dialkylzincs are often prepared and stored solvent-free. The first step in decomposing such compounds is to dissolve them under nitrogen in about 20 volumes of dry toluene, heptane, or some other hydrocarbon, which makes them less likely to ignite and easier to handle.

194 *Chemical Safety Matters*

A solution of the organometallic (less than 5% concentration) in a hydrocarbon or an ether is placed in a three-necked flask equipped with a stirrer, dropping funnel, nitrogen inlet, and ice bath. A 10% excess of t-butyl alcohol dissolved in a hydrocarbon is added dropwise to the well-stirred solution under nitrogen. This is followed by the addition of cold water and then enough 5% hydrochloric acid to neutralize the aqueous phase. The aqueous and organic phases are separated and disposed of appropriately. Where necessary, as for example with a dialkylcadmium, the metal ion can be separated from the aqueous phase by the procedure outlined in Section 10.2.2.

10.3.2. Metal carbonyls

Metal carbonyls are highly toxic materials, and many are sensitive to air. Because the metals are generally in a low oxidation state, the carbonyls can be destroyed by oxidation, in most cases by the gradual addition of the carbonyl to 25% excess of a well stirred chlorate(I) solution. The equipment and procedure used in the chlorate(I) oxidation of thiols (Section 10.1.6) are appropriate, but with the reaction being carried out under nitrogen. The metal carbonyl should be added as 5% solution in an inert solvent such as tetrahydrofuran or a hydrocarbon e.g.:

$$Ni(CO)_4 + OCl^- + H_2O \longrightarrow Ni^{2+} + Cl^- + 2OH^- + 4CO$$

10.3.3. Non-metal alkyls and aryls

Non-metal alkyls and aryls, such as BR_3, PR_3, and AsR_3, are sensitive to oxidation by air and other oxidizing agents, although the aryls are less sensitive than the alkyls. Many of these compounds are highly toxic. As with organic sulfides (Section 10.1.7), most of these compounds can be oxidized by 25% excess chlorate(I) (laundry bleach) to the corresponding oxides, which, since they are not pyrophoric, can be landfilled.

The alkyl non-metal hydrides, such as R_2PH and $RAsH_2$, can generally be decomposed by this chlorate(I) procedure. However, a few are so readily oxidized that they react violently with air, and for them the copper(II) sulfate oxidation procedure described for non-metal hydrides (Section 10.2.7) should be used.

10.3.4. Organomercury compounds

Organomercury compounds, such as R_2Hg and $RHgX$, must not be incinerated because some mercury may be volatilized. In many cases they can be oxidized as a dispersion in 25% excess laundry bleach [5.25% sodium chlorate(I)].

$$R_2Hg + 2OCl^- + H_2O \longrightarrow HgO + 2ROH + 2Cl^-$$

This reaction may be slow if the organomercury compound has low solubility in water. In such a case, the addition of a solution of bromine in carbon tetrachloride to a solution or suspension of the organomercury compound in carbon tetrachloride will result in oxidation to mercury(II) bromide, which can be separated and sent for recovery or to a landfill.

10.3.5. Calcium carbide

This reagent is no longer common in the laboratory, but is occasionally used as a source of acetylene. Small amounts of water in contact with calcium carbide can generate explosive acetylene/air mixtures that can be ignited by locally over-heated carbide. The following procedure for decomposing small amounts with aqueous acid minimizes this hazard:

$$CaC_2 + 2HCl \longrightarrow CaCl_2 + C_2H_2$$

Fifty grams of calcium carbide are suspended in 600 cm^3 of a hydrocarbon such as toluene or cyclohexane in a 2-dm^3 three-necked flask equipped with an ice bath, stirrer, dropping funnel, nitrogen inlet, and a gas outlet leading through plastic tubing to the back of the hood. With a moderate flow of nitrogen passing through the flask to carry off the acetylene generated, 300 cm^3 of 6 mol/dm^3 hydrochloric acid are added dropwise over a period of about 5 hours, and the mixture is stirred for an additional hour. The aqueous and hydrocarbon layers are separated and each disposed of appropriately. The aqueous layer can be flushed down the drain after being neutralized.

10.4. CLASSIFICATION OF UNLABELLED CHEMICALS FOR DISPOSAL

A laboratory chemical in a container without a label must be characterized as to general type (e.g., aromatic hydrocarbon, chlorinated solvent, mineral acid, soluble metal salt) so that it can be disposed of safely. It will generally suffice to carry out the preliminary steps of the systematic identification of organic compounds which are outlined below. The procedure is designed for organic substances, but most of the steps and principles apply also to inorganic substances. These can also be used to classify disposal of an "orphan" reaction mixture, i.e., an unlabelled material of unknown origin, or generated by a former worker in the laboratory.

When examining an unknown compound, it should be borne in mind that several classes of compounds are explosive or can form explosive peroxides on long exposure to air (Chapter 7). Accordingly, one should work cautiously behind a shield.

If the chemical is in a reagent container, it should be examined for any clue to the identity of the supplier, which may be revealing. In the case of an orphan reaction mixture, knowledge of the area of chemistry in which the worker was engaged may provide useful information as to chemical type. The size of the container may be informative. A large vessel is likely to hold a common chemical. Observe the colour, whether the substance is solid or liquid, and whether it is fluid or viscous. Test for acidity or basicity. Cautiously sniff the odour of the cap.

Place approximately 0.1 g of the substance in a small porcelain dish or a metal spoon. Bring a small flame near to see if the substance burns and, if so, how readily and with what appearance (for example, a sooty yellow flame suggests the presence of an aromatic ring). Heat it gently, then strongly. This gives an indication of volatility. If the substance is a solid, does it melt or decompose? Is there a residue? If so, it is probably an inorganic salt or oxide. Add a drop of water to see whether it dissolves and, if so, whether the solution is acidic or basic to pH paper.

Place a small sample of the original substance in a loop of a clean copper wire and hold it in a flame; halogenated compounds impart a distinct green colour to the flame.

Determine the solubility of the substance in water and, as appropriate, in ether, dilute sodium hydroxide, dilute hydrochloric acid, and concentrated sulfuric acid. The "solubility class" is very informative. If the substance is insoluble in any solvent tested, note whether it floats or sinks; this gives an indication of its density.

These steps can be carried out in a few minutes by an experienced chemist and often provide an adequate classification of the unknown chemical without further investigation. If more information is needed, the next step may be to obtain an infrared spectrum, which can indicate functional groups and whether the substance is a single entity or a complex mixture. Additional investigation can involve some of the simpler "classification tests", such as the ease with which it is oxidized with dilute aqueous potassium manganate(VII) or testing for loosely bound halogen with alcoholic silver nitrate. A nuclear magnetic resonance spectrum can often settle classifications that are still doubtful after all the above steps.

Some of the common inorganic oxidizing agents have characteristic colours, e.g., chromates(VI) are yellow, manganates(VII) are purple. Starch-iodide paper or other test paper (Section 7.2.1) can be used to indicate organic or inorganic oxidizing agents.

If the chemical is inorganic, is not pyrophoric, water-reactive, or a strong oxidizing agent, and does not liberate hydrogen cyanide or hydrogen sulfide on treatment with acid, it is generally not necessary to identify it precisely or to treat it before disposal in a secure landfill. Should it be necessary to identify the cation or anion, because of potential toxic characteristics, formal procedures should be used.

The results of the kinds of tests outlined above will generally provide sufficient information about an unknown chemical for assignment of its hazard class and selection of a safe method of disposal.

10.5. DISPOSAL OF LEAKING OR UNIDENTIFIED GAS CYLINDERS

Occasionally, a cylinder or one of its component parts develops a leak. Most such leaks occur at the top of the cylinder in areas such as the valve threads, safety device, valve stem, and valve outlet.

If a leak is suspected, do not use a flame for detection, but a commercial gas-leak detector or soapy water or other suitable solution. If the leak cannot be remedied by tightening a valve gland or a packing nut, emergency action procedures should be effected. Laboratory workers should never attempt to repair a leak at the valve threads or safety device, but should consult with the supplier for instructions.

The cylinder should be moved to an isolated area where it can be vented cautiously if the gas is flammable or oxidizing. If it is corrosive or toxic, the gas can be slowly released through a suitable neutralizer or detoxifier. If it is necessary to move a leaking cylinder through populated portions of a building, place a strong plastic bag, rubber shroud, or similar device over the top and tape it to the cylinder to confine the leaking gas. When the nature of the gas or the size of the leak constitutes a serious hazard, self-contained breathing apparatus (Section 2.4.2) and protective apparel (Section 2.3) may be required.

Cylinders with unknown contents pose particularly difficult problems and a considerable hazard. If the supplier can be identified, a phone call with a description of the cylinder may serve to identify the gas. If that fails, it is usually necessary to call in a waste-disposal firm that is capable of dealing with the problem.

11

Disposal of Explosives from Laboratories

An explosive chemical or mixture of chemicals is one that can undergo violent or explosive decomposition under appropriate conditions of reaction or initiation. Some chemicals used in laboratories are known to be explosive, and some combinations can be explosive (Appendix D). Laboratory manipulations with known explosive chemicals or reagent combinations should be carried out only by personnel who are thoroughly familiar with the hazards involved, the precautions that need to be taken (Sections 7.2, 7.3), and procedures for destroying or disposing of potentially explosive materials. Any laboratory procedure that results in an unexpected explosion should be investigated to ascertain the probable cause, and a laboratory safety rule established to prevent a recurrence. It is recommended that the circumstances of an unexpected explosion be submitted for publication in a widely read journal.

Explosive materials must be disposed of in a way that protects all personnel from the consequences of an explosion that might occur during the handling of the material. Potentially explosive material must not be disposed of in landfills, even in a lab pack. Small quantities of commercial explosives can be incinerated after reducing the explosive potential by dilution with a flammable solvent or a solid such as sawdust. Small containers of such diluted explosives should be fed into the incinerator one at a time. The incinerator operator should be fully aware of the nature of the materials being so handled. In general, however, this option is open only to laboratories that have access to their own incinerator and have experts on the handling of explosive materials. Potentially explosive materials should only be transported on public roads with specialized handling equipment and adequate protection.

One option for disposal of potentially explosive laboratory material is to arrange to have it detonated under carefully controlled conditions.

Disposal of explosives

Some laboratories may have personnel with training in handling explosives and the proper equipment, and may be able to remove and detonate the material on their site where no damage will result. Alternatively, some contract waste-disposal firms have the capability for removing and disposing of explosive material. A third possibility is to make arrangements with a local bomb squad or fire department to collect, remove, and detonate the material under safe conditions. In all these situations, the chemist should provide the disposal expert with whatever information is available on the hazards of the chemical.

Small quantities of some classes of explosive laboratory chemicals can be destroyed by procedures given in Chapter 10. However, there are members of each of these classes, as indicated in the following paragraphs, for which laboratory destruction procedures are not applicable. These exceptional compounds should be destroyed using detonation by personnel trained in handling explosives.

Many poly(nitro)aromatic compounds are explosive, and their disposal require the services of an expert trained in their handling (Section 10.1.14).

Picric acid (2,4,6-trinitrophenol) is normally sold in a damp condition, containing 10-15% water, and in this state it is relatively safe to handle. However, dry picric acid may explode on initiation by friction, shock, or sudden heating. Moreover, picric acid forms salts on contact with metals, and heavy metal picrates are highly sensitive to detonation by friction, shock, or heat. It is possible that some of the older reported explosions of picric acid were initiated by detonation of a minute quantity of a metal picrate in the threads of a metal-capped container. Although picric acid is now sold in plastic-capped containers, it is possible for material in such containers to dry out after repeated opening and thus become hazardous. If picric acid in a plastic-capped container appears to have dried out, the bottle can be immersed upside down in water for a few hours to allow water to wet the threads. The bottle can then be uncapped and filled with water. The water-filled bottle should be allowed to stand a few days to ensure complete wetting of the contents. Gram quantities of picric acid can be destroyed by reduction with tin and hydrochloric acid (Section 10.1.14). Larger quantities will require a commercial waste-disposal service or the local bomb squad or fire department. A metal-capped container of picric acid should be handled only by a trained expert, such as a member of a bomb squad.

Organic peroxides and hydroperoxides, including peroxide-containing solvents, can be treated as described in Section 10.1.16. However, any solvent capable of forming peroxides from which a solid has crystallized (e.g. aged diisopropyl ether) is potentially very hazardous and should be dealt with by personnel trained in handling explosives.

Most diazonium salts are not explosive and can be converted into disposable material by the coupling procedure described in Section 10.1.17. However, some diazonium salts are explosive when dry and should be carefully moistened before any manipulation is attempted.

Chloric(VII) acid can be disposed of as described in Section 10.2.3. However, ammonium chlorate(VII), heavy metal chlorates(VII), and chlorates(VII) of organic cations, e.g., pyridinium chlorate(VII), 2,4,6-trimethylpyrylium chlorate(VII), are inherently explosive and should be disposed of as explosives. Alkyl chlorates(VII) are also explosive, and their formation should be avoided. If one is formed inadvertently, it should never be isolated, and the mixture in which it exists should be disposed of as an explosive.

Sodium azide can be destroyed by the procedure described in Section 10.2.3. but heavy metal azides are too dangerous to be handled by this procedure and should be treated as explosives.

Sodium amide and potassium amide can be destroyed by the procedure of Section 10.2.3. However, as described in that Section, many other compounds in which nitrogen is linked to a metal should be disposed of as potential explosives.

12

A Waste-management System for Laboratories

Waste-management systems for a single laboratory operation, for a small college, and for a large university or industrial complex must have substantial differences in detail, but each system should have in common certain basic characteristics. Four elements that are essential to any laboratory waste-management system are:

1. Commitment of the laboratory chief executive to the principles and practice of good waste-management.
2. A waste-management plan.
3. Assigned responsibility for the waste-management system.
4. Policies and practices directed to reducing the volume of waste generated in the laboratory.

It is important that personnel at all levels - department heads, supervisors, academic faculty - exhibit a sincere and open interest in the waste-management plan. This support must be continuous. It is not sufficient to support the plan at its outset and to assume that it will then operate. The success depends on the participation and cooperation of the laboratory workers, who will be conditioned by their perception of management commitment. A programme that is perceived as having only nominal support will come to be ignored by laboratory personnel.

12.1. WASTE-MANAGEMENT PLAN

It is the responsibility of laboratory management to see that a plan is developed for the handling of surplus and waste chemicals. The plan should be tailored to the operations of the laboratory and should conform to all pertinent legal regulations. Written policies and procedures should be prepared to cover all phases of waste handling, from generation to ultimate safe and environmentally acceptable disposal. Although the necessary documents should be organized and written by the individuals who will be responsible for their implementation, there is much to be gained by having laboratory personnel participate in the planning process.

The documents should describe all aspects of the system for the particular laboratory and should spell out responsibilities and specific procedures to be carried out by laboratory personnel, supervisors, management, and the waste-management organization. The waste-management plan should be reviewed at regular intervals to be sure that it covers changes that may have occurred in the laboratory operations.

12.2. RESPONSIBILITY FOR THE WASTE-MANAGEMENT SYSTEM

A successful waste-management system requires a team effort from laboratory managers, laboratory supervisors, laboratory personnel, stockroom personnel, the local waste-management organization and the local safety and health organization.

The responsibilities of each of these groups should be set forth clearly in the waste-management plan. Although small laboratories may not have organizational subdivisions that correspond to each of these functions, the functions do exist, and the people who carry them out should assume the responsibilities assigned by the waste-management plan.

Responsibility for implementing the system must be specifically assigned. It is important that the individual in charge has enough knowledge of chemistry and laboratory work to be able to understand the problems faced by laboratory personnel in their part of the system and to make chemically sound judgments about particular waste-handling situations. If a waste manager does not have a broad chemical background, a consultant with appropriate chemical knowledge should be employed. Laboratory personnel are in the best position to identify any of their chemicals that might pose unusual hazards. They are responsible for putting waste in proper containers in accumulation sites. Laboratory supervisors must monitor laboratory operations and waste accumulation sites to check on proper performance under the plan. The responsibilities of stockroom personnel will depend on the size and complexity of the laboratory. Their activities may include dating the labels on chemicals whose use is time limited, monitoring stocks of such chemicals, and operating an exchange clearing house for un-needed chemicals (Chapter 9).

Even though these responsibilities in the operation of the waste-management system are made clear in the plan, they should be reinforced by training and refresher sessions.

12.3. REDUCTION OF VOLUME OF WASTE

Policies and practices for reducing the volume of waste generated in the laboratory and for avoiding special disposal problems should be an integral part of the waste-management plan. Some examples are given in the following sections.

12.3.1. Planning experiments

The planning of every experiment should include consideration of the hazardous properties of the starting materials; the disposal of leftover starting materials; the disposal of the products and by-products that will be generated. Questions to be considered include:

1. Are chemicals being acquired in proper quantities? Can any material be recovered for re-use? (Chapter 9)
2. Will the experiment generate any chemical that should be destroyed by a special laboratory procedure? If so, what procedure? (Chapter 10)
3. Can any unusual disposal problem be anticipated? If so, inform the waste-management organization beforehand.

12.3.2. Control of reagents that can deteriorate

Indefinite and uncontrolled accumulation of excess reagents can create storage problems and safety hazards. These problems can be alleviated, and purchase costs saved, by having an excess chemicals storage room to which laboratory workers can go for chemicals instead of ordering new material (Section 9.3). Chemicals that deteriorate with time can pose difficult disposal problems if allowed to accumulate in the laboratory. The waste-management system should provide for periodic searching for the following types of chemicals.

Reagents that react readily with oxygen or water deteriorate when stored for long times after the original container has been opened. Unnecessary or deteriorated samples of water-reactive chemicals (Appendix A) and pyrophoric chemicals (Appendix C) should not be allowed to remain in the laboratory. Severe hazards can be created by peroxide-forming chemicals (Appendix B) that have not been dated after opening the original container or that have exceeded the storage time limit after opening. Procedures for laboratory destruction of all three of these classes of chemicals are given in Chapter 10.

12.3.3. Maintenance of reagent labelling

Deterioration of labels is a common occurrence on old reagent containers. If the reagent itself or the container has not deteriorated, the container should be relabelled if its identity is certain. However, if a reagent container label has disappeared or become illegible and the identity of the chemical is unknown, the chemical should be discarded. The method for proper disposal can usually be determined by simple laboratory tests (Section 10.4).

12.3.4. Avoiding orphan reaction mixtures

Laboratory glassware containing reaction mixtures of unknown nature, and sometimes of unknown origin, can pose difficult disposal problems. Such materials occur frequently in research laboratories, particularly in those that have a high rate of personnel turnover. Simple laboratory tests (Section 10.4) may provide enough information for safe disposal of the material. The waste-management system should provide a procedure designed to prevent the occurrence of such orphan wastes.

Laboratories should require that any reaction mixture stored in glassware be labelled with its chemical composition, its date of formation, the name of the laboratory worker responsible, and a notebook reference. This procedure can provide the information necessary to guide the disposal of the mixture if the responsible laboratory worker is no longer available. It should be recognized, however, that such a procedure must be enforced and that it cannot guarantee that unlabelled mixtures will not be left behind by a departing worker.

12.4. THE WASTE-MANAGEMENT SYSTEM

The waste-management system for any laboratory should be tailored to the:

1. Volume and variety of wastes generated.
2. Number and locations of generating sources within the laboratory.
3. Options chosen for the disposal of wastes. These parameters should be determined as the system is being set up.

12.4.1. Initial survey

The first two of these parameters require a survey of the entire laboratory by the waste-system manager. For small laboratories, a brief survey that includes discussion with the individual responsible for each segment of the operation may suffice. For large, diverse departments it may be desirable to precede the survey with a simple questionnaire. It is almost inevitable that such a survey will uncover waste problems that must be solved at the outset, for example, caches of deteriorated or old chemicals, or forgotten chemicals of unknown origin or identity. The system must arrange for their safe disposal (Section 10.4) and should be designed to prevent such accumulations in the future.

The survey should include all units of the institution. Some laboratories will find units that are using chemicals and generating hazardous waste without the awareness of any hazard. This situation is particularly likely in academic institutions, where chemicals are used routinely in many areas, such as the biology, geology, electrical engineering, art, and physics departments, and in hospitals where workers may have little or no training in chemistry.

The survey should be designed to reveal the volume and types of non-hazardous waste being generated; the types of hazardous wastes being generated, and in what locations.

This information will determine many characteristics of the waste-management system, such as procedures for classification, segregation, and collection (Chapter 13), and the methods to be chosen for disposal of various types of hazardous wastes.

12.4.2. Waste-management manual

The waste-disposal plan should be summarized in a manual that is understandable by and available to all laboratory personnel. New laboratory workers should be made familiar with this manual by being given a copy at the time of their arrival or through an early induction programme.

The manual can be made more readable by putting long lists in appendices. Examples are:

1. A glossary of terms and abbreviations.
2. Listings of chemicals (as in Appendices A to E) such as: incompatible chemicals; potentially explosive chemicals and reagent combinations; water-reactive chemicals; pyrophoric chemicals; and peroxide-forming chemicals, including time limits on retention after opening the original container.

13

Identification, Classification, and Segregation of Laboratory Waste

A sound waste-management system requires that laboratory wastes be properly identified, classified as to the types of hazards they present, and segregated to avoid chemical interactions. These steps are essential for the safe accumulation, transportation, and disposal of the wastes. Moreover, in many countries, government regulations prescribe the classification and segregation of such wastes. The classification and segregation procedures for the laboratory should be set out clearly in the laboratory waste-management plan and summarized in the waste-management manual (Chapter 12).

13.1. IDENTIFICATION

The laboratory worker must decide if a material is no longer required and ought to be disposed of. Possibilities for recovery or recycling of the material (Chapter 9) should be considered. The laboratory operating guidelines should include information on the types of chemicals that can be recovered, recycled, or re-used. Once a material is declared to be waste, those working with it should determine the degree of hazard it represents and provide sufficient information on its characteristics for correct disposal. Accordingly, they must be familiar with the hazard characteristics by which wastes are classified and with the procedures used by the laboratory for segregating and collecting wastes. The laboratory waste-management manual should provide enough information for this purpose.

13.2. CLASSIFICATION

Wastes should be classified according to the type and degree of hazard, if necessary as prescribed by government regulation. The following guidelines should be helpful in classifying hazardous materials.

Acute hazardous wastes are substances that are fatal to humans in low doses, or that have an oral LD_{50} toxicity (rat) of less than 50 mg/kg, an inhalation LD_{50} toxicity (rat) of less than 2 mg/L, or a dermal LD_{50} toxicity (rabbit) of less than 200 mg/kg, or that are capable of causing serious, irreversible, or incapacitating illness.

Hazardous wastes are substances that meet any of the following criteria:

1. Flammability.
   (a) Liquids, other than aqueous solutions containing less than 24% (v/v) alcohol, that have a flash point below 60 °C.
   (b) Materials other than liquids that are capable, under ambient conditions of temperature and pressure, of causing fire by friction, absorption of moisture, or spontaneous chemical changes and, which when ignited, burn so vigorously and persistently as to create a hazard.
   (c) Flammable compressed gases that form flammable mixtures at a concentration of 13% (v/v) or less in air or that have a flammable range in air wider than 12% (v/v) regardless of the lower limit.
   (d) Oxidizing agents such as chlorate(VII), manganate(VII), or nitrate(V) or inorganic peroxides that readily yield oxygen to stimulate the combustion of organic matter.
2. Corrosivity, e.g. aqueous solutions that have a pH less than 2 or greater than 12.5.
3. Reactivity. This classification includes substances that react with water violently or to produce toxic gases or explosive mixtures; unstable or explosive substances; and substances that contain cyanide or sulfide or generate toxic gases when exposed to a pH in the range 2-12.5.

Classification and segregation of wastes by such criteria are essential for their safe handling and disposal. If the waste is not a common chemical with known characteristics, enough information about it must be supplied to satisfy regulatory requirements and to be certain that it can be handled and disposed of safely. For many wastes only the principal components need be specified. However, if the waste contains a carcinogen or a toxic metal, this information should be supplied. The information needed to characterize a waste also depends on the method of ultimate disposal.

13.3. SEGREGATION
13.3.1. Labelling

Classes of waste should be properly segregated for temporary accumulation and storage as well as for transportation and disposal. Accordingly, all wastes must be labelled properly before being removed. The label should contain sufficient information to assure safe handling and disposal, including the initial date of accumulation and chemical names of the principal components and any minor components that may be hazardous. The label should also indicate whether the waste is toxic, reactive, corrosive, metallic, flammable, an inhalation hazard, or is lachrymatory.

Some laboratories, particularly research laboratories, use and synthesize many unusual chemicals that can become waste. In general, chemists involved with such work know qualitatively whether they are flammable, corrosive, or reactive. If large quantities of chemical waste with unknown hazards are being generated, analytical tests must be carried out to determine such properies. However, for typically small quantities of laboratory chemicals, formal analysis is not warranted. Laboratory samples with unknown hazard characteristics, including orphan wastes (Section 10.4) can be tested on a small-scale in the laboratory for flammability with a small flame. The pH should also be determined, if the solution is aqueous, as should the reactivity with water or air. The presence of peroxides or other oxidizing compounds should be checked with potassium iodide (Section 7.2.1, 10.1.16). The one hazard characteristic that cannot be readily determined is toxicity, although the probability that a chemical is toxic can sometimes be inferred by analogy to closely related chemical structures. In the absence of basis for judgment, a waste should be assumed to be toxic and labelled accordingly.

Waste generated in the laboratory can often be characterized from knowledge of the starting materials, e.g., "hydrocarbon mixture", "flammable laboratory solvents", "chlorobenzene still bottoms". The professional expertise, common sense, judgment, and safety awareness of trained professionals performing chemical operations in the laboratory usually put them in a position to judge the type and degree of hazard of a chemical.

13.3.2. Disposal of wastes in sewer systems

Limited quantities of some wastes can be disposed of in sanitary sewer systems, but never in storm-sewer systems. A sanitary sewer is one that is connected directly to a water-treatment plant, whereas a storm sewer usually discharges into a stream, river, or lake. Guidelines for types and quantities of chemicals that can be disposed of in a sanitary sewer are given in Chapter 15.

13.3.3. Accumulating wastes for disposal

The first step in the disposal sequence usually involves accumulation or temporary storage of waste in or near the laboratory.

Except when a single chemical is being accumulated for recovery, waste accumulation generally involves several chemicals in a container. *ONLY COMPATIBLE CHEMICALS SHOULD BE PUT IN ANY CONTAINER,*

Identification, classification and segregation of waste

WHETHER PACKAGED SEPARATELY OR MIXED. In this context, "compatible" implies the absence of chemical interaction. Guidelines on *INCOMPATIBLE CHEMICALS* are given in Appendix E. The two common practices for accumulation of wastes are: mixing compatible chemicals in a waste container; and accumulating small containers of compatible wastes in a larger outer container, e.g., a lab pack (see next paragraph). The method chosen and the scheme for segregating the wastes depend primarily on the intended mode of disposal.

If laboratory wastes are to be landfilled (Chapter 18), the most common method of packaging is the lab pack. The procedure for preparing a lab pack is described in Section 18.4. The waste-management plan and manual should include specific directions for preparing lab packs as well as the assigned responsibility for preparing them. On the other hand, if a contract disposal service that prepares the lab packs is used, the manual should give directions for segregating and labelling wastes in accordance with the contractor's requirements. The principal consideration in segregation of chemicals for landfill disposal is compatibility. The laboratory waste-management manual should contain guidelines on incompatible chemicals or a list of them (Appendix E). In addition, explosives and chemicals that present a reactivity hazard (except for cyanide- and sulfide-bearing reactive wastes) must not be put into a landfill.

The method chosen for segregation and accumulation of wastes destined for incineration depends on the design of the incinerator and its waste feed mechanism, which vary widely. Some incinerators can handle only bulk liquid wastes, whereas others accept solid or packaged wastes such as fibre packs and glass or plastic bottles; a few even accept steel cans or drums.

Incinerators that accept only liquid wastes either blend them with fuels or incinerate directly. In either case the disposer generally pumps the contents from the container. Small containers are less desirable than the standard 200-litre drum. Incinerators that accept solid waste generally incinerate without removing the waste from the container, avoiding the hazards of opening containers. Some facilities will accept a wide variety of containers, including individual bottles. Others prefer to accept wastes in fibre packs, which is a combustible version of the lab pack that is used in landfills.

Explosive compounds should not be put into fibre packs for incineration. Some operators will incinerate certain types of explosives by adding them to the incinerator feed in small quantities, preferably diluted with a flammable solvent or sawdust. Containers with more than 100 cm^3 of carbon disulfide produce an explosive hazard in incinerators and should

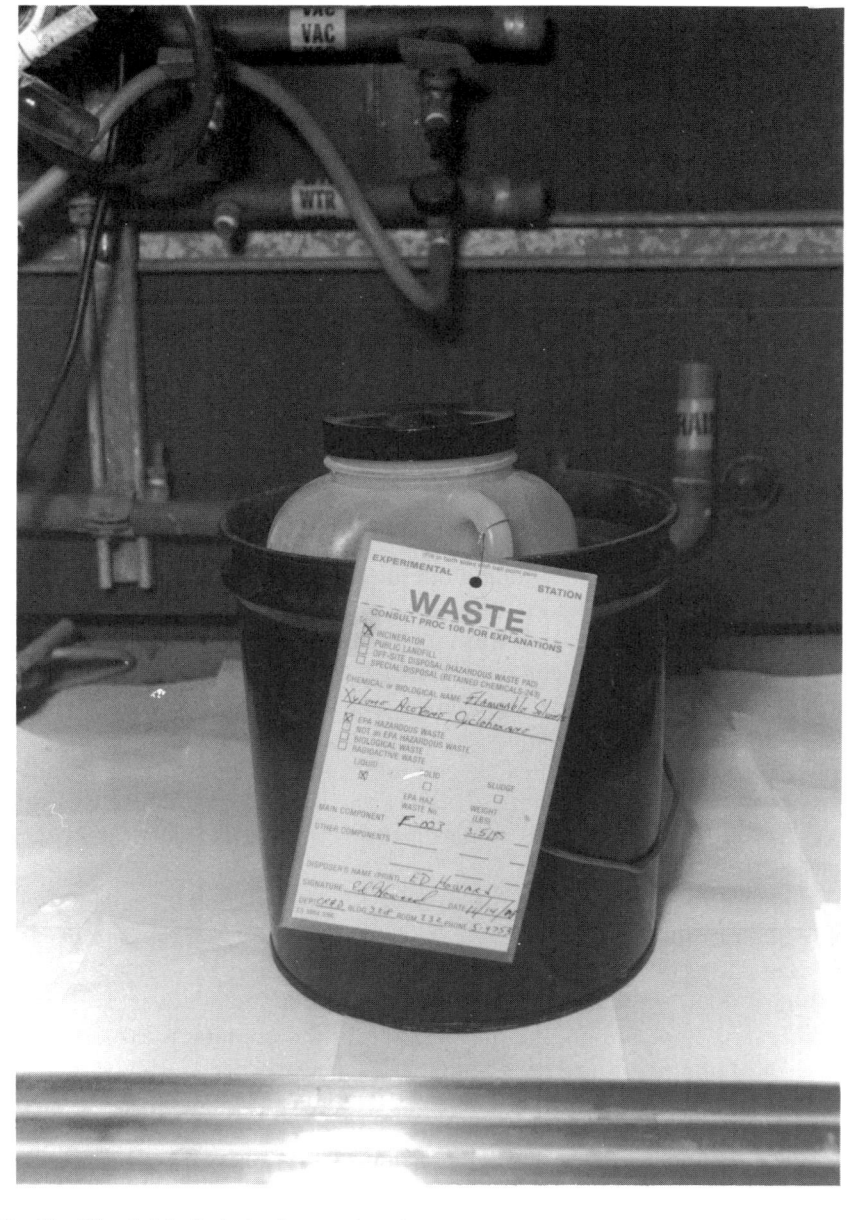

Fig. 22 Clearly labelled plastic container, in protective pail, for waste flammable solvents.

be avoided. A procedure for destruction of laboratory quantities of carbon disulfide is given in Section 10.1.7.

If separate incinerator facilities for hazardous and non-hazardous waste are available, segregation of hazardous waste can be both cost effective and environmentally more acceptable. Compared with incinerators for municipal waste, incinerators for hazardous waste require a more expensive design, more careful operation, higher costs for obtaining permits, and regulated treatment of ash and scrubber water. Consequently, incineration of non-hazardous waste in hazardous waste facilities is costly.

The proper sorting of wastes destined for incineration is essential. Halogenated wastes should be kept separate from non-halogenated wastes because the former must be burned in an incinerator with a scrubber that greatly reduces emissions of volatile halogen compounds.

Accumulation of laboratory wastes for disposal must be carried out in accordance with the written waste-management plan, which should include all safety factors (chemical incompatibilities being particularly important), regulatory requirements, and factors specific to the disposer and disposal method. A training programme should be initiated and maintained to assure that all laboratory personnel understand the requirements, such as accumulation procedures, safety procedures, and record keeping. In general, the design of a waste-accumulation point should follow the principles given for a waste-storage area (Chapter 14). It is especially important that the design conforms to local fire codes. However, sites for accumulation of small quantities of waste over a period of a few days need not be as elaborate as larger waste-storage areas. Thus modest quantities of single-type wastes can be accumulated temporarily in containers kept in a hood. Such containers should be clearly dated and labelled for content.

Large quantities of flammable wastes should never be accumulated in a closed, unventilated room. A vapour explosion in such a room would be disastrous. An appropriate container for each category of waste in the waste-accumulation plan should always be available. The container must be clearly marked to indicate the type of waste it can contain and the hazards associated with this category. It should be dated to indicate when accumulation started (Figure 22).

The accumulation area should be inspected regularly. The frequency of inspection can depend on the level of activity and the degree of hazard but generally should be at least weekly. The inspection should include the following considerations:

1. Adherence to the accumulation plan.
2. Condition of containers.
3. Availability of containers.

4. Dates of containers.
5. Adequate records of container contents.
6. Operation of safety equipment.

Documentation should be kept as to the contents of each accumulation container unless the container is labelled to receive, for example, only a single type of waste. The specific data required will depend on the size and complexity of the laboratory operation.

Although the same principles apply to smaller laboratories with simpler or fewer wastes, such laboratories may be able to work with simpler rules and requirements. For example, in a teaching laboratory where a specific experiment is being carried out by students, it should suffice to collect the wastes in containers labelled for each waste and to record the total quantity and date. A similar system can be used where specific waste solvents are being accumulated for recovery. If the wastes are to go into a lab pack for landfill disposal, or into an analogous fibre pack for incineration, the accumulation record should include the information that is on the individual waste containers.

14

Storage of Laboratory Waste

Containers of waste chemicals collected from individual laboratories must often be placed in temporary storage before being treated, disposed of, or stored elsewhere. Temporary storage times should be kept as short as possible. The storage facility should be designed for total containment of any spillage with as little release to the environment as possible. The design of the store must comply with all local regulations, including fire codes.

14.1. PROTECTION FROM THE WEATHER

Waste containers should be stored in an area where they are protected from poor weather, which can cause deterioration of labels, tags, and other markings. Such deterioration can create a hazard and could necessitate re-analysis of container contents. Exposure of containers to direct sunlight must be avoided as heat will cause pressure build-up in containers with volatile liquid contents. Containers should be tightly sealed so that water or moist air will not be sucked in when they are cooled. The introduction of moisture can result in container corrosion or chemical reaction with the contents.

A properly designed storage facility should have a roof to protect waste from sunshine and moisture. Warning signs should be displayed, and walls or fences erected to protect against unauthorized entry. It must have proper ventilation and be accessible to fire appliances and other emergency equipment. It should not be located in an area of high work density, but should be close enough to be used without difficulty and to allow proper surveillance and security.

14.2. PRIMARY CONTAINMENT

Waste should be put into containers that will be secure for the envisaged storage period until final disposal. If the waste is being stored prior to transportation, the use of containers approved under the pertinent transportation regulations is recommended, and may be required.

14.3. SECONDARY CONTAINMENT

It is prudent to use secondary containment under storage areas to catch potential leaks, spillages, and rainfall or snow that becomes contaminated. The storage-area base must be impervious to the waste being stored and to water. It should be designed so that leakage and precipitation do not come in contact with the containers or at least so that such contact is minimal. This can be accomplished by sloping the base toward a sump or other collecting basin, or by elevating the containers above the base. The containment should have levees, and the total capacity should be at least 10% of the total volume stored or equal to that of the largest container, whichever is greater.

Any material collected in a secondary containment area must be identified before being removed for treatment or disposal. Sometimes this identification is obvious, as in the case of rupture of a labelled container. In other cases it may be necessary to perform chemical tests before prescribing disposition of the material.

14.4. FIRE SUPPRESSION

The fire protection required for a storage area is often dictated by laboratory policies or by local codes and regulations. A storage area for flammable materials should have emergency fire equipment, including at least an appropriate fire extinguisher and breathing equipment. If large volumes of flammable materials are stored or if the store is near ignitable materials or structures, a sprinkler system is advisable.

14.5. INSPECTION

An area used for waste storage should be inspected regularly, preferably every week. A protocol should be followed that includes inspection of all containers, the safety and emergency equipment, and general housekeeping. Containers should be inspected for leaks, corrosion, proper closure, labels, and segregation of containers of incompatible content. Safety and emergency equipment should be inspected for general condition and expiration of service dates, and should be tested by operation if possible. Secondary containment should be checked and emptied if necessary. Any problems noted at inspection must be reported at once to the person responsible for the area so that they can be corrected promptly.

Inspections are more effective if a checklist is used and if the inspection is the only purpose of the visit. Inspections should not take place while putting in, or taking out, waste from the area. It is bad practice to consider inspections unnecessary because of frequent visits to the area for other purposes.

14.6. SEPARATION OF INCOMPATIBLE WASTES

It is essential to separate incompatible materials (Appendix E). It is particularly important to keep flammable materials away from oxidizing agents or sources of ignition. No incompatible materials should be put in the same container. Separation of their containers is not always required, although it is desirable. Additional requirements are often dictated by the ultimate route of disposal.

14.7. RECORD KEEPING

The record-keeping requirements for transportation, treatment, and disposal may be dictated by regulations. If not, it is still necessary to keep an inventory of wastes and contents of waste containers in storage. This may be done by establishing a log-in/log-out procedure for the storage area. It is poor practice to store containers whose contents are not known. Important information about waste holdings and correct disposal should not be kept in the immediate vicinity of the area concerned because it might be needed by emergency services in the event of some accident.

14.8. TRAINING

All personnel who use a waste storage facility should be thoroughly familiar with emergency equipment, emergency procedures, required inspections, waste-inventory procedures and rules for keeping incompatible wastes apart. Inadequate training can lead to serious problems.

14.9. PLANNING FOR EMERGENCIES

A storage area for hazardous waste should have plans for foreseeable emergencies. The plans should be reviewed periodically, and drills conducted. They should identify responsible persons who can coordinate any emergency response. They should be developed in consultation with, and submitted to, the local police and fire departments, hospitals, and response teams who might be called upon to provide emergency services.

14.10. CLOSURE OF A WASTE-STORAGE FACILITY

Any facility must be properly cleaned and, if necessary, decontaminated after it is taken out of active service. This includes disposal of any remaining waste, clean-up of secondary containment, and disposal of any contaminated equipment or releases.

15

Disposal of Chemicals in the Sanitary Sewer System

A few years ago it was common practice to dispose of most laboratory wastes down the laboratory drain. Today the indiscriminate disposal of chemicals, without regard to quantity or type, is not acceptable. Laboratory drains are connected to sanitary sewer systems, and their effluent will eventually arrive at a sewage-treatment plant. Some chemicals can interfere with the proper functioning of sewage-treatment facilities. In the laboratory drain system itself, some chemicals can create hazards of fire, explosion, or local air pollution; others can corrode the system.

It is essential to recognize that the characteristics and capabilities of waste water-treatment plants vary from one locality to another and that these factors and local regulations govern what types and concentrations of chemicals they can handle. The drain-disposal procedures and rules in the laboratory waste-management plan should be based on a thorough knowledge of local regulations pertaining to materials that are acceptable for disposal in the local sanitary sewer system. It may be helpful to provide the local waste water-treatment plant with information on the types and quantities of chemicals that the laboratory site plans to put into the sanitary sewer system and on the toxicities and biodegradabilities of these chemicals.

Subject to local regulations, certain types of chemicals are permissible for drain disposal while others are not. Therefore small quantities of some common laboratory chemicals can be safely and acceptably disposed of down the laboratory drain with proper precautions:

1. Only drains that flow into a waste water-treatment plant should be used. Systems such as a storm-sewer system that flow directly into surface water should never be used.

2. The quantities of chemicals disposed of in the drain must be limited to a few hundred grams or milliliters. Disposal should be performed by flushing with at least 100-fold excess water at the sink, so that the chemicals become highly diluted in the waste water effluent from the laboratory and even more so by the time the effluent reaches the treatment plant.
3. Sewer disposal of laboratory wastes by individual laboratory workers or by students in teaching laboratories should be monitored for adherence to the guidelines, with particular reference to types of chemicals, quantities, rates, and flushing procedures. Periodic checks by laboratory supervisors and by waste-management personnel are advisable. It is strongly advised, and may be mandatory in some localities, that the laboratory performs automatic sampling of its sanitary sewer effluent and that laboratory personnel collect and analyse daily composite samples of effluent. This is important for controlling drain disposal within limits required by local regulations.

An alternative procedure is to segregate, at the point of collection, those wastes that are suitable for sewer disposal, either as such or after some treatment. These collected wastes can then be fed at a low, controlled rate into the sewer system by a trained, authorized individual. The feed rate should be such that no interference with the sewage treatment facilities will occur, and should be determined by prior consultation with the management of these facilities.

15.1. DRAIN DISPOSAL OF CHEMICALS
15.1.1. Organic chemicals

Only those organic compounds that are reasonably soluble in water are suitable for drain disposal. A suitable definition is that used for qualitative organic analysis: a compound is considered water-soluble if it dissolves to the extent of at least 3%, i.e., whether 0.1 cm^3 or 0.1 g dissolves in 3 cm^3 of water in a test-tube. In general, such materials can be put down the drain, flushed or mixed with at least 100 volumes of excess water. However, this generalization must be tempered with chemical commonsense. Substances that boil below 50 $^{\circ}$C, even though adequately water-soluble, should not be poured down the drain because they can cause unacceptable concentrations of vapour. Diethyl ether, for example, may create a fire or explosion hazard. On the other hand, formaldehyde (b.p. -21 $^{\circ}$C) is an exception because it is so hydrated in water that very little vaporizes from dilute aqueous solution. Highly malodorous substances must not be put down the drain.

In general, a substance that meets the solubility criterion, but contains another material that does not, should not be poured down the drain. However, if the water-insoluble material comprises less than about 2% of the mixture, drain disposal is usually acceptable because the small quantity of water-insoluble material will be well dispersed in the aqueous effluent. Common examples are acetone that has been used to rinse glassware and ethanol that contains a hydrocarbon denaturant.

Some organometallic compounds such as Grignard reagents and alkyllithiums or aryllithiums can be decomposed by water into harmless solutions that are suitable for drain disposal. This decomposition should always be performed in the laboratory before drain disposal (Section 10.3.1) and never in the drain itself. The solubility criterion excludes hydrocarbons, halogenated hydrocarbons, nitro compounds, mercaptans, and most oxygenated compounds that contain more than five carbon atoms from drain disposal. Other exclusions are explosives, such as azides or peroxides, or water-soluble polymers that could form intractable gels. No more than unavoidable traces of highly toxic organic chemicals, whether of synthetic or biological origin, should be allowed into the system.

Guidelines on the types of organic compounds that are suitable for drain disposal are given in Appendix G. In general, compounds that are not listed in Appendix G are not suitable for drain disposal.

15.1.2. Inorganic chemicals

Drain disposal is permissible for dilute solutions of inorganic salts in which both cation and anion are listed (as being relatively nontoxic) in the right-hand columns of Tables 10.1 and 10.2. Salts in which either the cation or anion is listed in the left-hand columns are not permitted in more than unavoidable traces. Mineral acids and strong alkalies should be neutralized before drain disposal.

15.2. BIOLOGICAL PRETREATMENT OF LABORATORY WASTES

Some laboratory wastes that are not acceptable in the sanitary sewer system because of low solubility or chemical type can be biodegraded into material that is acceptable.

This biodegradation may be carried out at a local industrial waste water-treatment plant. The laboratory waste manager should find out if any local industries have their own pretreatment or treatment facilities, and if they are treating waste streams with similar wastes to those being generated in the laboratory. Because of the relatively small quantity of the laboratory waste, the industry may be willing to accept it into their treatment process. Such a practice has been used by some universities, which deliver the waste to the industrial treatment plant, where it is slowly added to the industrial waste stream.

An alternative is to set up an "in-house", laboratory-scale biological treatment plant. This option may be useful in laboratories in academic institutions, where knowledge applicable to the design and operation of such a plant exists, and where its operation can be a useful part of the teaching programme in addition to reducing the volume of hazardous waste. A small-scale biological treatment plant can pretreat many types of organic waste, making them acceptable for sanitary sewer disposal. Examples include phenols, alcohols, aldehydes, ketones, and waste from life science laboratories.

Many types of biological treatment processes have been developed by municipalities and industries, and these can be duplicated on a small scale in the laboratory. The treatment system can be either aerobic or anaerobic, although most waste water systems are aerobic, effecting oxidative degradation of organic substances. These are generally activated sludge or trickling filter systems. Bacteria for an aerobic pretreatment system can be obtained from the local waste water-treatment plant and gradually acclimatized to live on and degrade a specific waste. The pretreatment facility can be set up on a scale commensurate with the volume of waste to be treated, for example, in a 22-dm^3 round-bottomed flask, a fish aquarium, or a livestock watering tank. Such reactors can be operated on either a batch or a continuous basis. The rate at which they can biodegrade wastes depends on the chemical nature of the waste and the degree to which the bacteria are acclimatized to this type of chemical.

Anaerobic biodegradation is generally slower but can be useful for certain wastes that are not readily degraded in aerobic systems. Examples include aromatic hydrocarbons and their chlorinated derivatives. Laboratory-scale anaerobic digesters can be constructed from materials as simple as a 20-dm^3 carboy wrapped with heating tape and insulation.

16

Incineration of Hazardous Chemicals

An incinerator can be defined as an enclosed device using controlled flame combustion, the primary purpose of which is to break down hazardous waste thermally. This definition differentiates incineration from combustion of waste primarily for the recovery of its thermal value.

From an environmental point of view, incineration is probably the method of choice for the destruction of virtually all organic compounds as well as for some wastes that contain inorganic substances. In this Chapter, destruction by combustion of chemicals is discussed: in an off-site commercial hazardous waste incinerator; in a laboratory's own on-site hazardous waste incinerator; and as minor constituents of the fuel feed to a power- or steam-generating plant.

The combustion of laboratory wastes in municipal incinerators is generally impractical because these facilities are rarely equipped for, nor do they have permits for, burning of hazardous materials. The advantages of incineration, compared with disposal in a secure landfill, include:

1. Wastes are converted into harmless products, provided that there are proper emission controls.
2. There is no commitment to long-term containment of hazardous materials, as there is with a landfill.
3. Release of contaminants as a consequence of malfunction can be corrected relatively quickly and inexpensively. Malfunction of a secure landfill (e.g., leakage into groundwater) can usually be detected less readily and is more costly to remedy.
4. Incinerators handle most reactive wastes prohibited from landfills.

On the other hand, incineration has some disadvantages relative to disposal in a secure landfill:

1. A costly test-burn procedure may be required to obtain a permit for a hazardous-waste incinerator.
2. Incineration equipment is relatively expensive to install and maintain. As a result, incineration is more costly to the user than landfill disposal.
3. Incinerator emissions to the atmosphere must be controlled, recognizing that even carbon dioxide is a harmful pollutant.
4. A secure landfill (Chapter 18) is generally more accessible to laboratories than is a commercial incinerator. Many commercial incinerator operators will not accept laboratory waste because of its chemical diversity, relatively low volume, and disproportionate administrative costs.
5. Ash from a hazardous-waste incinerator must be disposed of in a secure landfill.

16.1. CLASSIFICATION OF WASTES FOR INCINERATION

Wastes destined for incineration have to be classified and segregated according to chemical type (Chapter 13). In addition, incineration characteristics depend on the physical state of the waste, so that classification into organic liquids, aqueous solutions and slurries, oil sludges and biosludges, and solids is necessary. Reputable incinerator operators also require information on the following: the chemical identity or class, unusual hazards, approximate heat of combustion, approximate ash content, and chemical compatibility with other wastes.

16.2. CRITERIA FOR WASTE INCINERATORS

Most wastes can be destroyed completely (efficiency 99.99%) by high-temperature oxidation. Some oxidation products, such as SO_x, NO_x, HCl, and metal oxides, may be noxious. Most of the SO_x is SO_2, with smaller amounts of SO_3. Nitrogen oxides are formed by oxidation of organically bound nitrogen in the waste, by decomposition of inorganic nitrates, and by thermal fixation of atmospheric nitrogen. Halogens in the waste produce a mixture of the hydrogen halide and gaseous halogen. Chlorine-containing compounds produce predominantly HCl instead of chlorine at incineration temperatures. Metal derivatives, depending on their volatility, may form solid residues or vaporize and re-condense to form an aerosol. Emissions of the latter may be acceptable, depending on the types and quantities of metal ions they contain, or they may be reduced to acceptable levels by appropriate air-pollution control devices.

Incinerators (Figure 23) must be designed to provide the conditions necessary to achieve complete oxidation, e.g., a supply of air in excess of the stoicheiometric requirement, adequate mixing of air and waste, and

Fig. 23 Incinerator showing the size and complexity of the plant needed for the safe processing of hazardous laboratory wastes.

sufficiently high temperature to complete oxidation in the residence time available.

16.2.1. Mixing requirements

Most small-scale incinerators are used for the disposal of liquid and solid wastes. Incinerators designed for injection of liquids must be able to atomize liquids to a droplet size of less than 150 μm diameter, so that the time for vaporization is short relative to the residence time in the combustion chamber. Solids in small units are usually supported by either a hearth or a grate. On heating, the solids release volatile products and leave a carbonaceous residue that is subsequently oxidized by reaction with oxygen.

In order to avoid local oxygen starvation, effective mixing must be achieved of the combustible vapours produced by vaporization of liquids or pyrolysis of solids. In addition, sufficient mixing energy and time must be provided to break down large-scale eddies to a small enough size for mixing to occur at a molecular level. The gross mixing is achieved by matching the injection of waste and air to provide a rather uniform distribution of both in the combustion chamber, by using high injection velocities to ensure high turbulence, and by the use of baffles and cross-jets to promote mixing. Air in excess of the stoicheiometric requirement can compensate for imperfect mixing.

16.2.2. Temperature requirements

The high-temperature oxidation of waste proceeds through a multistep, free-radical process. The many factors that control the rate of oxidation are not completely understood. The rates increase rapidly with increasing temperature. Oxidation may be inhibited by free-radical scavengers such as halogens. Temperatures for the gas phase reaction range from 1000 °C up to the adiabatic flame temperature, which may be over 2000 °C.

16.2.3. Time requirements

The time required for complete reaction decreases with increasing temperature, from approximately 2 s at 1000 °C to 50 ms at 2000 °C. However, the short reaction times attainable at higher temperatures are often precluded by the problems encountered with low-cost refractory materials above 1300 °C or with high-alumina refractories above 1500 °C.

16.2.4. Recommended operating conditions

The selection of conditions for incinerator operation can be indicated by laboratory-scale pyrolysis and oxidation. Several experimental reactors have been developed for determining the ease of destruction or incineration of chemicals, and have been used to screen chemicals for

Table 16.1. Calculated temperatures (°C) for destruction of selected compounds with efficiencies of 99% and 99.99% for residence times of 1 or 2 seconds

Compound	99%[a] T(°C)	Destruction efficiency 99.99%[a] Temperature T(°C)	99.99%[b] T(°C)
Acrylonitrile	-	729	703
Allyl chloride	-	691	649
Benzene	-	732	717
Carbon tetrachloride	-	-	820
Chlorobenzene	-	764	744
Chloroform	620	-	-
Dichlorodiphenyltrichloroethane	480	-	-
2,7-Dichlorodibenzo-p-dioxin	840	-	-
1,2-Dichloroethane	-	742	720
Hexachlorobenzene	-	-	880
2,2′,4,4′,5,5′-Hexachlorobiphenyl	730	-	-
Methyl chloride	-	869	823
Toluene	-	727	701
Triethylamine	-	594	570
Vinyl acetate	-	662	629
Vinyl chloride	-	743	722

[a] Residence time 1 s.
[b] Residence time 2 s.

trial burn selection. The chemicals to be tested are vaporized, premixed with excess air, and passed through a quartz tube heated to a known temperature in an electric furnace. The destruction efficiency, measured by analysis of the product gases, can be determined as a function of temperature and time spent in the reactor. Tests with such reactors indicate that common organic hydrocarbons, chlorinated organic compounds, nitriles, and amines are at least 99.99% destroyed within 2 s at temperatures of 629-880 °C (Table 16.1). These results apply to an incinerator that depends on the thermal destruction of compounds. Differences in the relative ease of destruction would be found for incineration in a flame which generates oxygen and hydroxyl radicals. These increase the destruction rate. Chlorinated compounds are readily degraded thermally because of the weak carbon-chlorine bond but may be difficult to burn because the chlorine radicals inhibit flame reactions. In an incinerator, temperatures higher than 600-900 °C would require adjustment of the time needed to vaporize and mix the chemicals for optimum efficiency. It is difficult to apply the results of small-scale experiments to full-scale units, and the performance of incinerators is usually determined by direct measurement.

16.3. OFF-SITE INCINERATION

Laboratories that do not have their own incinerator may find it cost effective to arrange for off-site incineration by a contract waste-disposal service (Chapters 12, 17). Some contractors pack laboratory waste and arrange for its transportation and incineration.

The alternative to specialized site incineration is to contract with a general commercial incinerator operator who is willing to accept the laboratory's waste. In order to do this, the laboratory must identify and segregate its waste (Chapter 13), and arrange for transportation to the incinerator (Chapter 17).

16.4. USE OF EXISTING BOILERS

Some wastes, such as liquid hydrocarbons and their oxygenated derivatives, can be burned for heat recovery as secondary fuels in steam boilers. Such wastes are usually mixed directly with the primary fuel, preferably in low proportion in order not to effect a change in the boiler operation. Local regulations and/or the terms of a boiler's operating permit may restrict the burning of waste chemicals. The primary function of a boiler is to generate steam, and it is inadvisable to introduce into the fuel stream a chemical that has a low heat of combustion. A suggested lower limit for the heat of combustion of a waste to be burned in such a manner is 18600 kJ/kg. This criterion excludes: polyhalogenated compounds; many polynitroaromatics; or compounds that contain significant amounts of nitrogen, phosphorus, or sulfur. These classes of compounds are also undesirable because of the potential interactions of their combustion products with boiler components.

16.5. ON-SITE INCINERATION

The successful destruction of hazardous wastes by incineration has been demonstrated both in large-scale facilities, many of which have the capability of accepting a variety of wastes, and in small units designed to handle wastes of known and fairly constant composition. There are problems, both technical and institutional, that need to be addressed when considering an incinerator for laboratory waste consisting of relatively small amounts of a wide variety of chemicals. Some of these are discussed in this Section.

An important question in the handling of packaged wastes is whether it is preferable to burn them in their packages, and accept the need for destroying the packages, or to transfer them to storage containers with compatible compounds to take advantage of the handling of larger batches. Combustion of solvents is best accomplished by the continuous flow of liquid of constant composition from a holding tank. Safe procedures must be established to protect operators from exposure to hazardous chemicals during transfer of waste into holding containers and the blending with fuel oil. Chemically incompatible materials (Appendix E) must not be mixed in such operations.

Incinerators are usually operated within a relatively narrow temperature range, the lower level being established by the temperature needed to ensure adequate destruction of the waste, and the upper level by the materials of construction. A burner fired with auxiliary fuel can provide the energy both for preheating an incinerator and for maintaining its temperature when handling wastes of low heat content. The liquids should be fed at low rates to avoid rapid temperature changes and possibly explosive gas surges from volatile liquids. Tightly closed containers should not be introduced into incinerators because pressure build-up can cause them to explode.

Cold surfaces in an incinerator will quench combustion reactions and lead to the emission of unburned or partially burned materials. This is most likely to occur in small units, which have a high surface to volume ratio. Wall quenching can be minimized by hot refractory surfaces in the incinerator. Waste should not be fed into an incinerator during periods of start-up or shutdown, when the walls are below operating temperature.

It is important, particularly for small institutions with limited resources, that an incinerator of the appropriate scale and design is selected to meet the specialized needs of the laboratory. There is little documentation on experience with small-scale on-site incinerators for the disposal of hazardous wastes from laboratories.

17

Transportation of Hazardous Chemicals

Most laboratories are not able to treat or dispose of their waste on the laboratory site. Hazardous wastes must be transported to a recognized landfill, incinerator, or treatment site, which may be some distance away. Packaging must be adequate to withstand the physical stress and temperature change likely to be encountered in transportation.

The laboratory that generates waste must comply with a variety of regulations, as well as the rules of the transportation contractor and the disposal site. It is important that all information concerning the properties of the waste material are made available in order to ensure safe transportation and disposal.

The regulations that cover transportation of hazardous chemicals are extensive and complex. A laboratory that chooses to pack and arrange for transportation of its waste must become thoroughly familiar with such regulations and develop the expertise necessary to comply with them. Assistance in assuring compliance is available from regulatory agencies, transportation and waste-disposal firms, and consultants. Approved materials, containers, and appropriate labels for shipping are commercially available. Laboratories that do not choose to become involved in packing and shipping their own waste can usually contract with a waste-disposal firm to do the packing and arrange for transportation and disposal.

17.1. RECOMMENDED TRANSPORTATION PROCEDURES

There are several key elements in the safe and legal transportation of waste, all of which should be included in the laboratory waste-management plan (Chapter 12).

17.1.1. Hazardous waste coordinator

It is essential to designate a hazardous-waste coordinator (Section 12.2), who has primary responsibility for implementing and overseeing all activities related to transportation of hazardous waste from the laboratory.

17.1.2. Regulatory requirements

The coordinator must be familiar with the regulations that control transportation of hazardous waste.

17.1.3. Selection of transportation and disposal contractors

Waste generators can be held liable for improper transport and disposal of their waste by others and should therefore investigate the companies with whom they contract for such activities. Apart from considerations of cost, the following points should be considered:

1. Competence to handle waste.
2. Financial stability.
3. Possession of appropriate permits.
4. Access to disposal facilities.
5. Method and site of disposal.
6. Knowledge and use of pertinent labels and documentation.
7. Education and experience of personnel.
8. Liability insurance coverage.
9. Waste storage capability.
10. References from recent users.

The hazardous-waste coordinator of the laboratory should not only be assured on these points before entering into a contract with a waste-disposal service but should also check on them periodically during the term of the contract. The coordinator should visit the disposal facility to observe its operation.

17.1.4. Designation of waste categories

The coordinator must establish waste categories for the laboratory, based on safe practices, and the regulations for disposal.

17.1.5. Packaging of waste

Hazardous waste material must be packaged properly for transportation. All containers must be strong and leakproof. Inner containers must be cushioned to prevent damage from impact during handling. Outer containers must be properly labelled and marked (Figure 24 and 25).

Small containers of laboratory waste are often packaged for transport in a lab pack, i.e., surrounded by an inert filler in an outer steel drum (Section 18.4). This pack is generally used for landfill disposal. An analogous pack in an outer fibre drum can be used for transportation to a waste-treatment facility or for incineration. If such a pack is intended for

Fig. 24 Hazardous-waste chemicals packed into a steel drum with inert absorbent filling material between the containers.

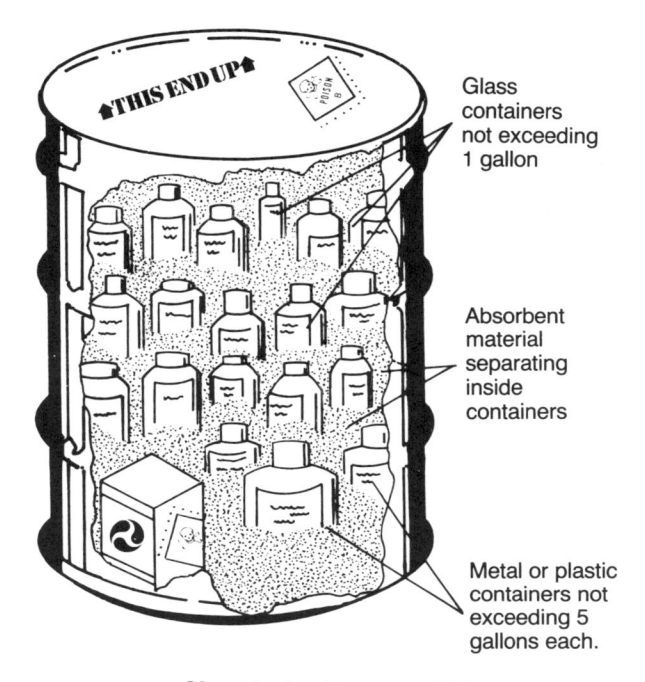

**Chemicals all compatible
and from one hazard class**

Fig. 25 Correct filling of a "lab pack" - a packaging unit officially recognized by the U.S. Department of Transport (DOT). This is the most appropriate way to ship waste laboratory chemicals, as it allows different materials from the same class of hazard to be packaged together. The treatment or disposal method to be used will determine the type of outer package (metal, fibre or plastic drum) and packing materials. (Reproduced, with permission, from a publication of the Task Force on the U.S. Resource Conservation and Recovery Act, American Chemical Society, Washington D.C., 1986.)

incineration, the incinerator operator may require that the inner individual containers are made of linear, high-density polyethylene.

17.1.6. Storage prior to transportation

Good practices and regulatory requirements for the storage of waste prior to transportation were outlined in Chapter 14.

17.1.7. Loading for shipment

It is important that the hazardous waste coordinator checks the transporter's vehicle for any obvious contamination and visible defects. All drums or other containers should be re-inspected prior to loading to ensure that they are not damaged. A final check should be made for appropriate labelling of waste containers. The coordinator should make certain that notices applicable to the hazard priority of the waste are provided.

Drums that contain liquids should be inspected to ensure there is no seepage at the seams or bungs. The contents of unacceptable drums should be transferred to acceptable drums, or the drum should be placed inside a recovery drum and packed with absorbent. During transfer operations, care should be taken to avoid inhalation of fumes or skin contact with chemicals.

17.1.8. Planning for emergencies

The hazardous-waste coordinator should have a response plan that anticipates possible emergencies such as a leaking drum in storage or in transport, a flash fire, or splashing of chemicals.

Appropriate protective clothing, footware, face protection, respiratory protection, and absorbent material for liquid spills should be readily available. To deal with the most likely incidents, personnel should be trained in techniques such as the plugging of pinhole leaks in large containers, spillage clean-up, and transfer of liquids from one container to another.

It is essential that vehicle drivers and disposal operators know of the hazards of the waste and are given instructions on how to react in emergencies. This is important, as experienced personnel may not be available during transportation and disposal.

18

Disposal of Hazardous Chemicals in Landfills

Waste that must be disposed of from the laboratory site is usually incinerated or put into a landfill. A sanitary landfill is one that is operated by a municipality or other agency for the disposal of household and municipal waste that is not considered hazardous. A secure landfill is one that has a permit from an appropriate governmental agency to receive and landfill waste that is defined by that agency as hazardous. In recent years, there have been significant changes in the general approach to the landfill disposal of laboratory wastes, and further changes are likely in the future. These changes involve increasingly strict limitations on what types of chemicals can be put legally into landfills. At one time, many chemical wastes were put into sanitary landfills intended for household and municipal waste, but now only wastes that are known to be non-hazardous may be put there, and this practice is subject to local regulations.

Many types of hazardous laboratory chemicals are collected and put into a secure landfill, but this situation is also changing. In many countries, increasingly strict regulatory requirements on the containment capability of secure landfills have resulted in the closing of a number of existing landfills and made the construction of new ones expensive. Accordingly, space in secure landfills is decreasing, and costs of disposal in them are increasing. In addition, disposal of chemicals in secure landfills poses the problem of long-term liability to the waste generator if the landfill containment fails. It is likely that laboratories wishing to construct their own secure landfills for disposal of their relatively small quantities of wastes may find that the costs are unacceptable.

The advantages and disadvantages of disposal in a secure landfill, compared to incineration, have been listed in Chapter 16. Although incineration is environmentally preferable, facilities are often not easily accessible. A secure landfill may appear to be the only practical option.

However, the trend of current regulations indicates that additional restrictions on landfill disposal are likely. Proposals being discussed include banning landfill disposal of certain types of liquids, liquids in a lab pack (Section 18.4) and possibly lab packs altogether. Accordingly, laboratories should limit the types and amounts of wastes they send to landfills and explore alternative routes of disposal. Ways of reducing waste volume are discussed in Section 12.3 and Chapters 9 and 10.

18.1. SANITARY LANDFILLS FOR NON-HAZARDOUS LABORATORY WASTES

Some solid chemical wastes produced in laboratories are not hazardous. While it may be expedient to treat these materials in the same manner as hazardous wastes, it is important to reduce the volume as much as possible. Non-hazardous wastes may be disposed of in a sanitary landfill if local regulations permit. However, these vary widely, and it is essential to consult local agencies to determine what types of wastes they will permit. If sanitary landfill disposal is available, the laboratory waste-disposal protocol should recommend separate containers, marked "Non-hazardous Waste," to ensure the safety of the waste handlers. Table 18.1 lists examples of non-hazardous wastes that have sufficiently low toxicity for safe disposal in a sanitary landfill. All the materials in this list have LD_{50} values greater than 500 mg/kg. Other criteria for toxicity may be used, depending on national regulations. The laboratory waste-disposal plan should list those materials in common use that are not hazardous.

18.2. DETOXIFICATION OF HAZARDOUS WASTES FOR LANDFILL DISPOSAL

Some hazardous laboratory wastes can be converted into non-hazardous wastes by treatment in the laboratory, resulting in residues suitable for disposal in a sanitary landfill. This course of action reduces the burden on a secure landfill, as well as the cost of using one, and should be pursued whenever the type and quantity of waste make it feasible. Procedures for laboratory destruction and/or detoxification of many types of common laboratory chemicals are given in Chapter 10.

18.3. COMMERCIAL SERVICES

There are commercial firms that provide contract services for disposal of laboratory wastes. Some of these firms have employees with sufficient knowledge and experience to be able to prepare lab packs (Section 18.4), and may supply their own packing materials. They arrange for transportation of the lab packs and disposal of the wastes. These firms will usually also deal with the administrative requirements for transportation and disposal. A laboratory that uses such a service must have a waste-collection and segregation system that conforms to the contractor's

Table 18.1. Examples of non-hazardous laboratory wastes

A. *Organic chemicals*

Sugars and sugar alcohols
Starch
Naturally occurring amino acids and salts
Citric acid and its Na, K, Mg, Ca, NH_4 salts
Lactic acid and its Na, K, Mg, Ca, NH_4 salts

B. *Inorganic chemicals*

Sulfates: Na, K, Mg, Ca, Sr, Ba, NH_4
Phosphates: Na, K, Mg, Ca, Sr, NH_4
Carbonates: Na, K, Mg, Ca, Sr, Ba, NH_4
Oxides: B, Mg, Ca, Sr, Al, Si, Ti, Mn, Fe, Co, Cu, Zn
Chlorides: Na, K, Mg
Fluorides: Ca
Borates: Na, K, Mg, Ca

C. *Laboratory materials not contaminated with hazardous chemicals*

Chromatographic adsorbent
Glassware
Filter paper
Filter aids
Rubber and plastic protective clothing

---

requirements. It must be remembered however that the generator of the waste may be responsible for its safe and legal disposal, even when this is carried out by a contractor. Accordingly, a laboratory using such a service should investigate the firm thoroughly before signing a contract and should check regularly to ensure that operations continue to be carried out in a reliable manner (Section 17.1.3). It is sometimes possible to transport waste to a commercial treatment plant for eventual landfilling. However, such facilities are designed to handle large quantities of individual wastes from manufacturing plants by treating them chemically or physically to convert them into wastes that can be landfilled. Accordingly, they are often unwilling to deal with laboratory wastes because of the uneconomically small quantities and because of their need to know the composition of wastes to be treated. Commercial treatment plants generally require analysis of the wastes they receive. If a single drum contains a large number of chemical compounds, analysis of its contents is not practical, with the result that the wastes may not be acceptable. Some installations, however, will accept such a drum of waste on the basis of a list of the chemicals it is known to contain.

18.4. DISPOSAL OF HAZARDOUS WASTES IN A SECURE LANDFILL

The most common method of secure landfill disposal of laboratory wastes is the use of a lab pack (a 200-L open head steel drum) that is filled with small containers of chemicals packed in and separated by an absorbant medium. The requirements for a lab pack are summarized below.

The drums must be of the open head variety to allow the proper placement of the inside containers and absorbent. Inside containers must not react with, be decomposed by, or be ignited by the waste held in them, and must be non-leaking and tightly and securely sealed. The inside containers must be securely surrounded by enough absorbent material to completely absorb all the liquid contents of the inside containers and to fill the drum completely. The absorbent material must not be capable of reacting with, or being decomposed or ignited by, the contents of the inside containers. Vermiculite and Fuller's earth are commonly used for packing around containers of solids because of their cheapness, availability, and inert nature. Commercial absorbents, such as Oil Dry®, have greater absorptive capacity than vermiculite or Fuller's earth and are preferable for packing around containers of liquid.

Incompatible wastes (Appendix E) must not be placed in the same drum. The purpose of this restriction is to prevent potentially dangerous reactions between wastes packaged in the same lab pack. Wastes must not be landfilled if they can explode or release toxic gases when exposed to water, strong force, or heat.

There are strict requirements for labelling and paperwork for the transportation of any hazardous chemicals, including lab packs. The laboratory hazardous-waste coordinator must be thoroughly familiar with these requirements and be certain that they are followed (Section 18.3).

18.5. SOLIDIFICATION OF LIQUID WASTES BY ABSORPTION

Liquid wastes that are not in a lab pack can be put into a secure landfill only if they have been solidified by combining them with sufficient inert absorbent that no free liquid remains and then placed in a steel drum. The procedure can be useful for handling wastes that cannot be disposed of by incineration or other combustion and that are too large for lab pack disposal. A small-scale test should be made to determine that the absorption process is not strongly exothermic. The absorbents should be the same type as those used in lab packs (Section 18.4). This method cannot be used for any of the types of reactive materials that are prohibited from landfill disposal (Section 18.4) or for corrosive materials. Only compatible liquids (Appendix E) should be so treated for packaging in a single drum.

19

Disposal of Chemically Contaminated Waste from Life-science Laboratories

Life-science laboratories vary widely in the kinds of activities carried out, and hence in the kinds of waste they generate. This may be solid or liquid and includes such materials as carcasses, tissues, animal bedding and excrement, plant material, and microorganism cultures. Any of these may be contaminated with hazardous chemicals, and with microorganisms (e.g., bacteria, fungi, viruses) that may be pathogens.

Chemical operations are often carried out in life-science laboratories, and the disposal of this waste is already described elsewhere in this monograph. This chapter deals with biological waste with or without the presence of hazardous chemicals, although emphasis is placed on biological waste that is chemically contaminated. This is a category that, unlike purely chemical waste or purely biological waste, is scarcely considered in either the scientific literature or regulations.

Procedures for handling and disposal of biological waste, which range from the disinfection or sterilization of contaminated materials to the incineration of animal carcasses, are well documented. Specific and generally accepted protocols for the removal and disposal of biological wastes and their associated contaminated materials have evolved from years of experience and research in academic, industrial, and governmental laboratories and hospitals.

Reliance on autoclaving, incineration, or burial as a treatment or disposal strategy is prevalent, although in some cases inappropriate. While generally recognized as standard procedures, their efficacy is not universal and their application is not always appropriate. Thus heat-resistant spores, animal carcasses, or some biological toxins (e.g., aflatoxin) may not be effectively treated by autoclaving. Prudent, effective, and less-costly alternative handling techniques exist for innocuous as well as for hazardous biological wastes. Disinfectants that contain active chlorine or iodine are useful for killing bacteria, bacterial spores, viruses, rickettsiae, and fungi on large pieces of equipment subject to surface contamination or in heat-sensitive materials whose subsequent use or effectiveness might be compromised if autoclaved. It is the responsibility of investigators to characterize their biological wastes and judiciously apply known techniques, materials, and equipment to minimize the hazard to health and environment and the cost to the laboratory.

19.1. DEFINING THE WASTE

The first step in managing any waste is to characterize it. For chemically contaminated biological waste (CCBW), this step consists of identifying types and sources of the waste, determining its rate of generation, and describing its physical, chemical, and biological properties.

As already mentioned, typical examples of CCBW from life-science laboratories include dead animals and animal tissues, animal bedding and excrement, cell cultures, and body fluids. The chemical contaminant usually results from administration of chemicals to determine their effects on life systems or application of chemicals for a germicidal or disinfecting purpose. Usually, the mass of the chemical or drug is quite small in comparison with the mass of the biological substance. For example, an animal challenged with a suspected carcinogen might contain only microgram or milligram quantities of the challenge material. Similarly, plants used in agricultural research laboratories may be treated with minute amounts of insecticide or fungicide. Because the apparent physical properties of the combined CCBW are usually the physical properties of the biological material alone, it is quite common to manage CCBW in the same manner as purely biological waste is managed. Such a waste-management assumption is not always valid, as will be discussed further.

Before a treatment and/or disposal strategy can be selected, consideration must be given to the nature of the chemical contaminant associated with the biological material. The following questions must be considered before selecting the most effective treatment/disposal process:

1. What is the chemical contaminant, and how much was added to the biological material?
2. What is the fate of the contaminant within the biological system? How much chemical is metabolized or otherwise altered, and how much remains unchanged? Are more hazardous reaction products formed?
3. What are the known toxic properties of the chemical contaminants or reaction products?

4. Are the chemicals or reaction products easily volatilized, oxidized, or otherwise affected by the input of some form of energy?
5. Does the waste contain radioactive material? Disposal of such material is strictly regulated in most countries.

Although answers to the above questions will not always be readily available, they should be sought in order to minimize risks from the selection of an inappropriate treatment or disposal technique.

19.2. TREATMENT AND DISPOSAL TECHNIQUES

It is a basic principle that viable organisms in biological materials must be killed before they leave the laboratory. This is true even if their destination is an incinerator, for they may contaminate the ash or be carried off in the gases if incineration is not rapid and complete. This principle applies particularly to certain viruses and spore-forming organisms. For animal tissue and excrement containing their normal microbial population, direct incineration is usually acceptable. However, if the material may contain disease-producing organisms, autoclaving or other treatment should precede incineration. With proper selection of temperature, pressure, and time, autoclaving is often highly efficient, provided that any packages are open to access of steam. Treatment is usually achieved by heat/pressure, such as autoclaving, wet heat, dry heat; or by chemicals (wet/gaseous), such as chlorine, iodine, phenols, acidified alcohols, ethylene oxide, ozone, formaldehyde, or various quaternary ammonium salts.

Ultimate disposal, following treatment, is usually through the sewer system for liquids, or a sanitary landfill for solids. Treated products can include autoclaved tissue-culture fluids, ash and residue from a pathological-waste incinerator, and pipettes that have been soaked in a chemical disinfectant such as a chlorine solution. Regulations may control wastes discharged to both sewage systems and sanitary landfills, and consequently regulations at these levels should be consulted in determining the method of disposal.

If the biological waste contains a chemical contaminant, the selection of the method of treatment/disposal becomes more complicated than when considering only biological waste. For example, autoclaving a cell culture contaminated with a highly volatile carcinogen could lead to a dangerous release of the carcinogen. In incineration - a common treatment for dead animals - the volatilization of the chemical contaminant may be beneficial, particularly if the chemical itself is oxidized to harmless end products. However, if the contaminant contains a volatile metal, such as mercury or lead, or an organic compound that is difficult to destroy by combustion, such as tetrachlorodibenzodioxin, incineration may actually serve to disperse a toxic substance into the air, and some other method of disposal should be chosen. Similarly, for chemical decontamination of pathogenic materials that contain added chemicals, an assessment of the interaction between the chemical species is helpful in anticipating potentially harmful end products (e.g., chlorination yielding chloramines) and in assuring that the amount of disinfecting chemical is adequate to destroy the pathogens.

In some cases, considerations such as these will help to identify a treatment that satisfactorily addresses both chemical and biological problems. It is possible, however, that the chemical contaminant is in such high concentration or of such a nature that traditional biological treatment methods are not suitable. In such cases it may be necessary to give priority to the chemical aspect of the CCBW and treat it as a hazardous waste. Often the CCBW can be treated and rendered innocuous by chemical-waste incineration, remembering that multipurpose chemical-waste incinerators are usually operated under more rigorous combustion conditions than pathological incinerators (Chapter 16). Otherwise, disposal in a secure landfill may be the solution.

19.3. INCINERATION

Incineration, already widely used to dispose of biological wastes with or without chemical contamination, is growing in importance, hence some detail on its use is justified.

Incineration is a clean, efficient method of disposing of animal carcasses, organs, excrement and the paper or other bedding on which it is collected, and wood or cardboard containers in which animals have been shipped. The preferred type of incinerator for carcasses and organs is the pathological incinerator, which has a solid hearth rather than a grate in the ignition chamber. A disadvantage of the grate is that small carcasses and organs may drop through without being destroyed. Moreover, the hearth facilitates release of the water that makes up about 85% of animal tissue and must be evaporated before destruction of organic matter can take place. Location of primary and secondary burners in the incinerator is critical. Flames must impinge directly on the wastes to maximize evaporation and combustion. There are good reasons to have the incinerator on the laboratory site if local ordinances permit it, and if there is enough material to make the unit economical. The advantages are that direct control of the operation helps to ensure proper incineration.

The argument for incineration is stronger for animal material containing hazardous chemicals than for chemical-free material, as there are relatively few landfills approved for hazardous chemicals and these may involve high transport costs.

Wet animal waste, such as carcasses and excrement, should be placed in plastic bags, which should then be tied or taped. Bags that are to be incinerated on the site should be placed in a sealed cardboard carton or in a heavy laminated paper bag. Paper bags should be folded at the top and stapled. For ease of handling, loaded bags should not weigh more than 20 kg. If bags containing carcasses are not collected and incinerated daily, they should be stored frozen or in a walk-in refrigerator maintained at 4 °C.

Glass and metal should be excluded from bags going to a hearth-type incinerator because of the possibility of explosion of closed bottles or cans and increased incinerator maintenance due to fused glass or excessive ash. Appropriate labels should be attached, such as *ANIMAL CARCASSES* or *CAUTION-CANCER SUSPECT AGENT*.

A laboratory with access to good incineration facilities should not assume that all its biological waste can or should be incinerated. It is likely that some of the laboratory's biological waste could be treated by other methods for safety, environmental, or economic reasons. However, recognition that incineration is not always appropriate should not obscure the fact that it is an excellent way to dispose of most chemically contaminated biological waste.

Appendix A

Water-reactive Chemicals

This Appendix lists some common laboratory chemicals that react violently with water and that should always be stored and handled so that they do not come into contact with liquid water or water vapour. They must not be disposed of in landfills, even in a lab pack, because of their characteristic reactivities. Procedures for decomposing laboratory quantities are given in Chapter 10, with the pertinent section given in parentheses.

Alkali metals (10.2.4)
Alkali metal hydrides (10.2.3)
Alkali metal amides (10.2.3)
Metal alkyls, such as lithium alkyls and aluminium alkyls (10.3.1)
Grignard reagents (10.3.1)
Halides of non-metals, such as BCl_3, BF_3, PCl_3, PCl_5, $SiCl_4$, S_2Cl_2 (10.2.5)
Inorganic acid halides, such as $POCl_3$, $SOCl_2$, SO_2Cl_2 (10.2.5)
Anhydrous metal halides, such $AlCl_3$, $TiCl_4$, $ZrCl_4$, $SnCl_4$ (10.2.4)
Phosphorus(V) oxide (10.2.9)
Calcium carbide (10.3.5)
Organic acid halides and anhydrides of low molecular weight (10.1.10)

Appendix B

Peroxide-forming Chemicals

Many common laboratory chemicals can form peroxides in the presence of atmospheric oxygen. A single opening of a container to remove some of the contents can introduce enough air for peroxide formation to occur. Some types of compounds form peroxides that are violently explosive in concentrated solution or as solids. Accordingly, peroxide-containing liquids should never be evaporated to dryness. Peroxide formation can also occur in many polymerizable unsaturated compounds, and these can initiate a "runaway", sometimes explosive polymerization reaction. Procedures for testing for peroxides and for removing small amounts from laboratory chemicals are given in Section 10.1.16.

This Appendix provides a listing of structural characteristics of organic compounds that can peroxidize and of some common inorganic materials that form peroxides. Although the tabulation of organic structures may seem to include a large fraction of the common organic chemicals, they are listed in approximate order of decreasing hazard. Reports of serious incidents involving the last five organic structural types are extremely rare, but they are included because laboratory workers should be aware that they can form peroxides that can influence the course of experiments in which they are used.

This Appendix also provides specific examples of common chemicals that can become serious hazards because of peroxide formation. Suggested time limits are given for retention or testing of these compounds after opening the original container. Although some laboratories mark containers of such chemicals with the date of receipt of the original container, it should be recognized that such dating does not take into account the unknown time span between original packaging and the date of receipt. The date of opening the original container of a chemical that is likely to form peroxides should always be marked on the container. Labels such as that illustrated below should be provided to all laboratory workers to date and affix to samples of peroxide-forming reagents.

> **PEROXIDIZABLE COMPOUND**
> Date Received ....
> Date Opened ....
> Discard or test within 6 months after opening

The material in this Appendix has been adapted from Jackson *et al.*, (1974). This reference and Sections 7.2.1 and 10.1.16 should be consulted for additional information on labelling and handling of peroxide-forming chemicals.

Table B.1 gives examples of common laboratory chemicals that may form peroxides on exposure to air. The lists are not exhaustive, and analogous compounds that have any of the structural features given in Table B.2 should be tested for the presence of peroxides before being used as solvents or distilled. The recommended retention times begin with the date of synthesis or of opening the original container.

Table B.1. Types of chemicals that may form peroxides

Organic structures

    Ethers and acetals with alpha hydrogen atoms
    Olefins with allylic hydrogen atoms
    Chloroolefins and fluoroolefins
    Vinyl halides, esters, and ethers
    Dienes
    Vinylacetylenes with alpha hydrogen atoms
    Alkylacetylenes with alpha hydrogen atoms
    Alkylarenes that contain tertiary hydrogen atoms
    Alkanes and cycloalkanes that contain tertiary hydrogen atoms
    Acrylates and methacrylates
    Secondary alcohols
    Ketones that contain alpha hydrogen atoms
    Aldehydes
    Ureas, amides, and lactams that have a H atom linked to a C attached to a N.

Inorganic substances

    Alkali metals, especially potassium, rubidium, and cesium (Section 10.2.4)
    Metal amides (Section 10.2.3)
    Organometallic compounds with a metal atom bonded to carbon (Section 10.3)
    Metal alkoxides

Table B.2. Common peroxide-forming chemicals

Severe peroxide hazard on storage with exposure to air: *Discard within 3 months*

Diisopropyl ether	Sodium amide (sodamide)
Divinylacetylene[a]	Vinylidene chloride (1,1-dichloroethylene)[a]
Potassium metal	Potassium amide

Peroxide hazard on concentration: do not distil or evaporate without first testing for the presence of peroxides: *Discard or test for peroxides after 6 months*

Acetaldehyde diethyl acetal (acetal)	Ethylene glycol dimethyl ether (glyme)
Cumene (isopropylbenzene)	Ethylene glycol ether acetates
Cyclohexene	Ethylene glycol monoethers (cellosolves)
Cyclopentene	Furan
Decalin (decahydronaphthalene)	Methylacetylene
Diacetylene	Methylcyclopentane
Dicyclopentadiene	Methyl isobutyl ketone
Diethyl ether (ether)	Tetrahydrofuran
Diethylene glycol dimethyl ether (diglyme)	Tetralin (tetrahydronaphthalene)
Dioxan / Dioxolan	Vinyl ethers[a]

Hazard of rapid polymerization initiated by internally formed peroxides[a]

List A. Normal liquids: *Discard or test for peroxides after 6 months*[b]

Chloroprene (2-chloro-1,3-butadiene)[c]	Vinyl acetate
Styrene	Vinylpyridine

List B. Normal gases: *Discard after 12 months*[d]

Butadiene[c]	Vinylacetylene[c]
Tetrafluoroethylene[c]	Vinyl chloride

---

[a] Monomers may polymerize and should be stored with a polymerization inhibitor from which the monomer can be separated by distillation just before use.

[b] Although common acrylic monomers such as acrylonitrile, acrylic acid, ethyl acrylate, and methyl methacrylate can form peroxides, they have not been reported to develop hazardous levels in normal use and storage.

[c] The hazard from peroxide formation in these compounds is substantially greater when they are stored in the liquid phase. If stored in this form, without an inhibitor, they should be included in List A.

[d] Although air cannot enter a gas cylinder in which gases are stored under pressure, these gases are sometimes transferred from the original cylinder to another in the laboratory, and it is difficult to be sure that there is no residual air in the receiving cylinder. An inhibitor should be put into any secondary cylinder before transfer. The supplier can suggest an appropriate inhibitor to be used. The hazard posed by these gases is much greater if there is a liquid phase in the secondary container. Even inhibited gases that have been put into a secondary container under conditions that create a liquid phase should be discarded within 12 months.

Appendix C

Pyrophoric Chemicals

Many members of the following readily oxidized classes of common laboratory chemicals ignite spontaneously in air, therefore the following special precvautions are necessary.

Pyrophoric chemicals should be stored in tightly closed containers under an inert atmosphere (or, for some, an inert liquid), and all transfers and manipulations of them must be carried out under an inert atmosphere or liquid. Pyrophoric chemicals cannot be put into a landfill because of their reactivity. Suggested disposal procedures are outlined in Chapter 10 and given in parentheses after each class.

Grignard reagents, RMgX (10.3.1)

Metal alkyls and aryls, such as RLi, RNa, R_3Al, R_2Zn (10.3.1)

Metal carbonyls, such as $Ni(CO)_4$, $Fe(CO)_5$, $Co_2(CO)_8$ (10.3.2)

Alkali metals, such as Na, K (10.2.4)

Metal powders, such as Al, Co, Fe, Mg, Mn, Pd, Pt, Ti, Sn, Zn, Zr (10.2.4)

Metal hydrides, such as NaH, $LiAlH_4$ (10.2.3)

Non-metal hydrides, such as B_2H_6 and other boranes, PH_3, AsH_3 (10.2.7).

Non-metal alkyls, such as R_3B, R_3P, R_3As (10.3.3)

Phosphorus (white) (10.2.8)

A more extensive list that includes less-common chemicals can be found in Bretherick, (1990).

Appendix D

Potentially Explosive Chemicals Combinations

Table D.1 lists some common classes of laboratory chemicals that have the potential for producing a violent explosion when subjected to shock or friction. These chemicals should never be disposed of as such but should be handled by procedures suggested in Chapters 10 and 11. Information on these and some less common classes of explosives is available in Bretherick (1990).

Table D.1. Shock-sensitive compounds

Acetylenic compounds, especially polyacetylenes, haloacetylenes, and heavy metal salts of acetylenes (copper, silver, and mercury salts are particularly sensitive)
Acyl nitrates(V).
Alkyl nitrates(V), particularly polyol nitrates(V) such as nitrocellulose and nitroglycerine
Alkyl and acyl nitrites(III)
Alkyl chlorates(VII)
Amine metal oxosalts: metal compounds with coordinated ammonia, hydrazine, or similar nitrogenous donors and ionic chlorate(VII), nitrate(V), manganate(VII) or other oxidizing group
Azides, including metal, non-metal, and organic azides
Chlorate(III) salts of metals, such as $AgClO_2$ and $Hg(ClO_2)_2$
Chlorate(VII) salts. Most metal, non-metal, amine, and organic cation chlorates(VII) can be detonated / undergo violent reaction in contact with combustible materials
Diazo compounds such as CH_2N_2
Diazonium salts, when dry
Fulminates (Silver fulminate, AgCNO, can form in the reaction mixture from the Tollens' test for aldehydes if it is allowed to stand for some time. This can be prevented by adding dilute nitric(V) acid to the test mixture as soon as the test has been completed.)
Hydrogen peroxide becomes increasingly treacherous as the concentration rises above 30%, forming explosive mixtures with organic materials and decomposing violently in the presence of traces of transition metals
N-Halogen compounds such as difluoroamino compounds and halogen azides

Table D.1. (continued)

N-Nitro compounds such as *N*-nitromethylamine, nitrourea, nitroguanidine, and nitric amide
Oxo salts of nitrogenous bases: chlorates(VII), dichromates(VI), nitrates(V), iodates(V), chlorates(III), chlorates(V), and manganates(VII) of ammonia, amines, hydroxylamine, guanidine, etc.
Peroxides and hydroperoxides, organic (Section 10.1.16)
Peroxides (solid) that crystallize from or are left from evaporation of peroxidizable solvents (Section 10.1.16 and Appendix B)
Peroxides, transition-metal salts
Picrates, especially salts of transition and heavy metals, such as Ni, Pb, Hg, Cu, and Zn; picric acid is explosive but is less sensitive to shock or friction than its metal salts and is relatively safe as a water-wet paste (Chapter 11)
Polynitroalkyl compounds such as tetranitromethane and dinitroacetonitrile
Polynitroaromatic compounds, especially polynitro hydrocarbons, phenols, and amines

Table D.2 lists illustrative combinations of common laboratory reagents that can produce explosions when they are brought together or that give reaction products that can explode without any apparent external initiating action. This list is not exhaustive.

Table D.2. Potentially explosive combinations of some common reagents

Acetone with chloroform in the presence of base
Acetylene with copper, silver, mercury, or their salts
Ammonia (including aqueous solutions) with Cl_2, Br_2, or I_2
Carbon disulfide with sodium azide
Chlorine with an alcohol
Chloroform or carbon tetrachloride with powdered Al or Mg
Decolourizing carbon with an oxidizing agent
Diethyl ether with chlorine (including a chlorine atmosphere)
Dimethyl sulfoxide with an acyl halide, $SOCl_2$, or $POCl_3$ or with CrO_3
Ethanol with calcium chlorate(I) or silver nitrate(V)
Nitric(V) acid with acetic anhydride or acetic acid
Picric acid with a heavy-metal salt, such as of Pb, Hg, or Ag
Silver oxide with ammonia with ethanol
Sodium with a chlorinated hydrocarbon
Sodium chlorate(I) with an amine

Appendix E

Incompatible Chemicals

The term "incompatible chemicals" refers to chemicals that can react with each other violently; with evolution of heat; or to produce flammable products or toxic products. Incompatible chemicals must not be placed in the same lab pack for transport or landfill disposal and must always be handled, stored, and packed so that they cannot accidentally come into contact with each other. Guidelines for the segregation of common laboratory chemicals that are incompatible are presented in Tables E.1 and E.2. Table E.1 contains general classes of compounds that should be kept separated; Table E.2 lists specific compounds that can pose reactivity hazards. Chemicals in each grouping in columns A and B of each table should be kept separate. Further information on specific chemical reaction hazards can be found in Bretherick, (1986 & 1990).

Table E.1. General classes of incompatible chemicals

Column A Acids and oxidizing agents[a]	Column B Bases, metals and reducing agents[a]
Chlorates(VI)	Ammonia, anhydrous and aqueous
Chromates(VI)	Carbon
Chromium(VI) oxide	Metals
Halogens / Halogenating agents	Metal hydrides
Hydrogen peroxide	Nitrates(III)
Manganates(VII)	Organic compounds in general
Nitric(V) acid / Nitrates(V)	Phosphorus
Peroxides	Silicon
Sulfates(VII)	Sulfur

[a] The examples of oxidizing and reducing agents are illustrative of common laboratory chemicals. The listings are not intended to be exhaustive.

Table E.2. Specific chemical incompatibilities

Column A	Column B
Acetylene, monosubstituted acetylenes	Group IB and IIB metals and their salts
	Halogens / Halogenating agents
Ammonia, anhydrous and aqueous	Halogens / Halogenating agents
	Mercury
	Silver
Alkali and alkaline earth	Water
carbides	Acids
hydrides	Halogenated organic compounds
hydroxides	Halogenating agents
metals	Oxidizing agents[a]
oxides / peroxides	
Azides, inorganic	Acids
	Heavy metals and their salts
	Oxidizing agents[a]
Cyanides, inorganic	Acids / Strong bases
Mercury and its amalgams	Acetylene
	Ammonia, anhydrous and aqueous
	Nitric(V) acid
	Sodium azide
Nitrates(V), inorganic	Acids
	Reducing agents[a]
Nitric(V) acid	Bases
	Chromic(VI) acid
	Chromates(VI)
	Metals
	Manganates(VII)
	Reducing agents[a]
	Sulfides
	Sulfuric acid
Nitrates(III), inorganic	Acids
	Oxidizing agents[a]
Organic compounds	Oxidizing agents[a]
Organic acyl halides	Bases
	Organic hydroxy and amino compounds
Organic anhydrides	Bases
	Organic hydroxy and amino compounds
Organic halogen compounds	Group IA and IIA metals
	Aluminium
Organic nitro compounds	Strong bases
Oxalic acid	Mercury and its salts
	Silver and its salts
Phosphorus	Oxidizing agents[a]
	Oxygen
	Strong bases

Table E.2. (continued)

Column A	Column B
Phosphorus(V) pentoxide	Alcohols
	Strong bases
	Water
Sulfides, inorganic	Acids
Sulfuric acid (concentrated)	Bases
	Potassium manganate(VII)
	Water

[a] See list of examples in Table E.1.

Appendix F

pH Ranges for Precipitating Hydroxides of Cations

Most metal ions are precipitated as hydroxides or oxides at high pH. However, many precipitates will redissolve in excess base. For this reason, it is necessary to control pH closely in a number of cases. The Table shows the recommended pH range for precipitating many cations in their most common oxidation state, which will not dissolve in 1 mol/dm^3 sodium hydroxide (pH 14). This information is from Erdey (1965) and Burns et al., (1981).

Table F.1. *pH Range for precipitation of metal hydroxides and oxides*

Metal	pH range	Metal	pH range
Ag(I)	9 to 14	Pb(II)	7 to 8
Al(III)	7 to 8	Pd(II)	7 to 8
As(III)	Precipitated as sulfide	Pd(IV)	7 to 8
As(V)	Precipitated as sulfide	Pt(II)	7 to 8
Au(III)	7 to 8	Re(III)	6 to 14
Be(II)	7 to 8	Re(VII)	Precipitated as sulfide
Bi(III)	7 to 14	Rh(III)	7 to 8
Cd(II)	7 to 14	Ru(III)	7 to 14
Co(II)	8 to 14	Sb(III)	7 to 8
Cr(III)	7 to 14	Sb(V)	7 to 8
Cu(I)	9 to 14	Sc(III)	8 to 14
Cu(II)	7 to 14	Se(IV)	Precipitated as sulfide
Fe(II)	7 to 14	Se(VI)	Precipitated as sulfide
Fe(II)	7 to 14	Sn(II)	7 to 8
Ga(III)	7 to 8	Sn(IV)	7 to 8
Ge(IV)	6 to 8	Ta(V)	1 to 10
Hf(IV)	6 to 7	Te(IV)	Precipitated as sulfide
Hg(I)	8 to 14	Te(VI)	Precipitated as sulfide
Hg(II)	8 to 14	Th(VI)	6 to 14
In(III)	6 to 13	Ti(III)	8 to 14
Ir(IV)	6 to 8	Ti(IV)	8 to 14
Mg(II)	9 to 14	Tl(III)	9 to 14
Mn(II)	8 to 14	V(IV)	7 to 8
Mn(IV)	7 to 14	V(V)	7 to 8
Mo(VI)	Precipitate as Ca salt	W(VI)	Precipitated as Ca salt
Nb(V)	1 to 10	Zn(II)	7 to 8
Ni(II)	8 to 14	Zr(IV)	6 to 7
Os(IV)	7 to 8		

Appendix G

Guidelines for Disposal of Chemicals in the Sanitary Sewer

The following lists comprise compounds that are suitable for disposal down the drain with excess water in quantities up to about 100 g. However, local regulations may prohibit drain disposal of some compounds and these regulations should be checked before any laboratory compiles its list of compounds acceptable for disposal down its own drains. Compounds on both lists are water-soluble to at least 3% and present a low toxicity hazard. Those on the organic list are readily biodegradable.

Table G.1. Non-hazardous organic and inorganic chemicals suitable for sanitary sewer disposal

A. Organic chemicals

Alcohols
    Alkanols with fewer than 5 carbon atoms
    t-Amyl alcohol
    Alkanediols with fewer than 8 carbon atoms
    Glycerol
    Sugars and sugar alcohols
    Alkoxyalkanols with fewer than 7 carbon atoms
    n-$C_4H_9OCH_2CH_2OCH_2CH_2OH$
    2-Chloroethanol
Aldehydes
    Aliphatic aldehydes with fewer than 5 carbon atoms
Amides
    $RCONH_2$ and $RCONHR$ with fewer than 5 carbon atoms
    $RCONR_2$ with fewer than 11 carbon atoms
Amines[a]
    Aliphatic amines with fewer than 7 carbon atoms
    Aliphatic diamines with fewer than 7 carbon atoms

Table G.1. (continued)

Benzylamine
Pyridine
Carboxylic acids[a]
   Alkanoic acids with fewer than 6 carbon atomsa
   Alkanedioic acids with fewer than 6 carbon atoms
   Hydroxyalkanoic acids with fewer than 6 carbon atoms
   Aminoalkanoic acids with fewer than 7 carbon atoms
   Ammonium, sodium, and potassium salts of the above acid
      classes with fewer than 21 carbon atoms
   Chloroalkanedioic acids with fewer than 4 carbon atoms
Esters
   Esters with fewer than 5 carbon atoms
   Isopropyl acetate
Ethers
   Tetrahydrofuran
   Dioxolane
   Dioxane
Ketones
   Ketones with fewer than 6 carbon atoms
Nitriles
   Acetonitrile
   Propionitrile
Sulfonic acids
Sodium or potassium salts of most are acceptable

B. Inorganic chemicals[b]

Cations	Anions
Al(III)	BO_3^{3-}, $B_4O_7^{2-}$
Ca(II)	Br^-
Cu(II)	CO_3^{2-}
Fe(II), (III)	Cl^-
H	HSO_3^-
Li	OH^-
Mg	I^-
Na	NO_3^-
NH(IV)	PO_4^{3-}
Sn(II)	SO_4^{2-}
Sr	SCN^-
Ti(III), (IV)	
Zn(II)	
Zr(II)	

[a] Those amines and acids with a disagreeable odour, such as 1,4-butanediamine, dimethylamine, butyric acids, and valeric acids, should be neutralized, and the resulting salt solutions flushed down the drain, diluted with at least 1000 volumes of water.
[b] This list comprises water-soluble salts in which both the cation and anion have a low toxic hazard. Any of these salts that are strongly acidic or basic should be neutralized before being flushed down the drain.

Glossary

Acetylcholinesterase - Enzyme responsible for breaking down acetylcholine and likely to be inhibited by the organophosphorus pesticides and nerve-gas-like chemicals.

Activated carbon (charcoal) - Used to adsorb components of gas mixtures or impurities from liquids as they permeate across its surface.

Adiabatic flame - Combustion in which the flame temperature may be over 2000 °C.

Advance planning - Preventing or limiting problems with chemical waste-disposal by developing suitable guidelines, rules, etc. and implementing their use.

Aerosols - Liquid particles with diameter between 0.001 and 100 μm in a gas phase. The smaller the particle size, the greater the risk of inhalation.

Allergic reactions/responses - See *Allergy*.

Allergy - Pathological symptoms following exposure to a previously encountered substance (allergen) which would otherwise be classified as harmless; essentially a malfunction of the immune system.

Alumina-activated sieve - Adsorbent in pellet form used to purify gases and liquids.

Amalgamation - Alloying with mercury.

Anaerobic biodegradation - Bacterial and fungal degradation of products in the absence of air.

Anemometer (velometer) - A device used to measure airflow rates in open or enclosed areas and at the faces of input or exhaust ports.

Anoxia - Absence or relative lack of oxygen; reduction of oxygen in the body tissues to below physiological partial pressures.

Armoured hood - See *Barricade*.

Artificial resuscitation (respiration; pulmonary resuscitation) - Maintenance of breathing by mechanical or mouth-to-mouth ventilation in the absence of spontaneous lung function.

Ascarite® - Sodium hydroxide on inert material used in pellet form to adsorb carbon dioxide from air or other gas mixtures.

Ataxia - Failure of muscular coordination; irregularity of muscular action.

Auto-ignition temperature - See *Ignition temperature*.

Barricade (armoured hood) - Protective hood that will shield the operator against flames or explosive particles.

Bentonite clay - Highly adsorbent Fuller's earth (q.v.) clay used in particle form for containing chemical spills (available to the public as cat litter).

Biodegradation - Biodegraded - Breakdown of chemicals, plastics, polymers etc. by the reaction of light, heat, water, bacteria, fungi, each on its own or in concert.

Biological monitoring - The quantitation of human or animal exposure to a chemical based on the measurement of the parent compound or its metabolite(s) in biological materials or excreta, e.g. blood, urine and exhaled air.

Biological treatment plant - The use of biological systems involving bacteria or fungi under aerobic or anaerobic conditions to facilitate biodegradation of chemical wastes for safe disposal.

Biological waste - Materials that are generated by life science laboratories such as dead animals and animal tissues, animal bedding and excrement, body fluids, plant-derived material, cell cultures, etc. It may also be contaminated with chemicals, e.g. waste generated in animal or microbiological tests for studies of the toxicity or the mode of action of a chemical.

Bradycardia - Slowness of the heartbeat, as evidenced by the pulse rate falling to less than 60 per minute.

Bronchitis - Inflammation of bronchi (lung), which may be acute or chronic.

Bunsen tube - A pressure release device.

Cancer - The disease which results from the development of a malignant tumour and its spread into surrounding or distant tissues. See also *Tumour*.

Cat litter - See *Bentonite clay*.

Carboxyhaemoglobin - The molecule formed when carbon monoxide binds (reversibly) to haemoglobin, so blocking its oxygen-binding capability. This effect may lead to asphyxia, even when the inspired oxygen concentration is sufficient to support respiration.

Carcinogen - A chemical or biological agent that can act on living tissue in such a way as to cause a malignant neoplasm.

Carcinogenesis - The induction by chemical, biological or physical agents, of neoplasms that may or may not be observable; an earlier induction of neoplasms than are usually observed; the induction of more neoplasms than are usually observed. Fundamental differences in the underlying mechanisms may be involved in the different cases.

Cardiopulmonary resuscitation - A combination of cardiac massage (by compression of the chest wall) to restart the heartbeat, with artificial resuscitation (q.v.) to ensure blood oxygenation.

Cartridge respirators - A respirator that draws ambient air through (a) cartridge(s) of suitable adsorbent(s), e.g. activated charcoal, to remove harmful substances. Their efficacy and safety depends on the use of an appropriate cartridge that is active on a relatively low concentration of toxic substance, and a sufficiently high concentration of oxygen in the inspired air.

Catch pot - A container to retain liquid or slurry discharged from an overflow or pressure-relief device.

Celite® - Siliceous or diatomaceous earth (q.v.) which is used, like kieselghur or fire-brick, for its absorptive, porous or abrasive characteristics.

Chelation - The trapping of a multivalent ionic species (often a metal cation) by ionic bonding to a larger molecule. If used therapeutically, the action is intended to render the ion inactive in the biological matrix, and also to aid its excretion.

Chemical allergy - Allergy caused by a specific chemical or group of chemicals.

Chemical decontamination - The process of removing residual amounts of harmful chemicals from glassware or other equipment so as to facilitate their safe use or repair. The process should be based on a knowledge of the properties of the contaminant(s), so as to effectively remove or destroy such substances by appropriate chemical or physical treatment.

Chemical incompatibility - Any interaction of a chemical and an environmental factor (e.g. water, light, air), or between two or more chemicals, that may give rise to a chemical reaction or produce an enhanced hazard.

Chemically contaminated biological waste - Biological waste (q.v.) that contains pollutant or toxic chemicals.

Chemical reaction - A process in which a substance or mixture of substances is converted into one or more new substances.

Chemical spills prevention - A planned way of dealing with or preventing chemical spills, e.g. a systematic method for the containment of chemicals that includes small volumes of stock containers, closures that facilitate safe transfer of materials, break resistant containers, or working over trays lined with absorbent material.

Chronically toxic substance - A chemical that causes damage after exposure of repeated or long-duration, usually in low concentration.

Clean-up reporting - Action taken following a spillage or contamination which documents containment, disposal and decontamination procedures, together with suggestions to prevent similar accidents.

Cold trap - A device that is used to condense vapour or gas or to freeze liquid, so as to contain it, or prevent it from reaching parts of a system where it might do harm.

Colour coding - A system by which organizations or agents that sell gas cylinders (or other chemical containers) use a nationally or internationally recognized scheme for coloured labelling or marking. However, the codes may not be observed in filling the gas cylinders or containers, and the paint or label may peel off if there is inappropriate storage. Thus colour coding may not be a reliable method for identification of the contents; the name stencilled or printed on the cylinder or container is a better method of identification.

Commercial laundry bleach - Solution of sodium hypochlorite sold as a household bleach (e.g. Clorox®, which contains 5.25% w/v (0.75 mol/L) sodium hypochlorite). The trade names of such products and their nominal concentrations differ throughout the world.

Containment - To physically separate polluted equipment or waste or a contaminated atmosphere, from an area that is not to be contaminated.

Contaminant - In some contexts (e.g. in the use of gas-cleaning equipment) the word is used as a synonym for "pollutant".

Contract disposal service - A commercial organization that specializes in waste-disposal for a broad range of chemical substances by landfill, incineration or other means, according to strict local regulations.

Controlled area - An area within a laboratory or other workplace to which only suitably trained personnel have access and who must undertake work practices within a defined frame of reference.

Controlled flame combustion - The combustion of waste in conditions such that only fully oxidized volatile products are produced, together with ash.

Convulsion (epileptic seizure) - A violent spasm of neurological origin that may be precipitated by exposure to a chemical or physical stimulus.

Corrosive - A substance that destroys living tissues on contact, by a direct chemical action.

Critical temperature - The temperature of the liquid-vapour critical point (at a defined pressure for a given substance), above which the substance does not exist in liquid phase.

Crossed polarizers - The use of two plane polarized sources of light to view birefringent materials or to characterize crystalline substances.

Cryogenic materials - A broad range of substances used as coolants, that can freeze human tissue on contact, e.g. solid carbon dioxide or liquid nitrogen.

Cumulative poison - A substance that can produce tissue damage or pathologic effect through exposure to small concentrations with absorption over a long period of time, e.g. cadmium, absorbed from vapour produced in brazing operations.

Cytotoxic - Causing disturbance to cellular structure or function, often leading to cell death.

Decontamination - A means of rendering harmless by neutralization, elimination or removal, a toxic or potentially toxic substance in the laboratory or general environment. See also *Chemical decontamination*.

Developmental poison - Any toxic substance that adversely affects the sperm, ova, embryo or fetus.

Diaperene - A tetraalkylammonium salt, commonly used to deodorize toilets.

Diatomaceous earth - Fossil residues of diatoms that are used in a fine particulate form for their adsorptive properties. See also *Celite*®.

Disposable respirator - Respirator that is used once and discarded - generally of the activated charcoal filter type.

Disposal - Any form of treatment technology that converts, destroys or removes chemical waste in a safe manner.

Dry ice - Solid carbon dioxide, which is useful as a coolant down to about -75 °C. There are various trade names for this material.

Drying train - A device used to dry a flow of gas or liquid by chemical absorption.

Earthed/earthing - See *Grounded/grounding*.

Eddy current - Electric current induced within a conductor when it moves through a non-uniform magnetic field.

Emergency equipment - Any equipment, such as respirators, containment and clean-up kits, for specified emergency use (NOT FOR ANY OTHER USE) that is part of the *emergency plan* (q.v.); as such it must be checked and maintained regularly.

Emergency plan - A documented set of procedures that should identify evacuation routes and shelter areas, the location of medical facilities, and the proper means of dealing with and reporting all accidents and emergencies. The plan should be reinforced by periodic drills and simulated emergencies.

Equipment maintenance programme - The systematic checking of equipment to ensure its optimum functioning and maximum safety; suitable records must be kept by a responsible person.

Evacuation procedure - The logistics of how to leave a laboratory area in an emergency. It should be established, documented, communicated to all personnel, and practised in periodic drills or simulated emergencies.

Exothermic reaction - A chemical reaction that generates or liberates heat. Such a reaction may add to the risk of combustion or thermal burns in a chemical accident.

Explosive polymerization - An untoward consequence of an exothermic process associated with polymerization of a chemical or mixture of chemicals.

Exposure (in environmental health and toxicology) - The concentration (or intensity) of a particular chemical or physical agent that reaches the target organism, organ or body. Exposure is usually expressed in numerical terms of duration, frequency, and concentration (for chemical agents and microorganisms) or intensity (for physical agents such as ionizing radiation).

Failsafe design - A feature of the design or functioning characteristics of a system specifically incorporated to prevent dangerous malfunction, e.g. a power shutdown device if the coolant of a heated system is cut off.

Flammable - A chemical is said to be flammable (in preference to inflammable) if it readily catches fire and burns in air. A flammable liquid does not itself burn; it is the vapour evaporated from the liquid that burns. An important property of a flammable gas or vapour is described by two limits which define the range of concentrations in air that will ignite and may explode - the *upper* and *lower explosive (flammable) limits.* (q.v.).

Flammable-storage cupboard or container - A device specially designed for the safe storage of solvents or other flammable materials to avoid their combustion even if they are involved in a fire.

Flash point - The lowest temperature at which the vapour of a volatile solvent or oil will ignite in air, with a flash or spark.

Fibre pack - A leak-resistant container made of flammable material that is used to collect, store and transport chemicals before they are incinerated.

First-in -first-out system - A method used in chemicals storage and distribution so that there is least chance of accumulating old stocks that may deteriorate or build up hazardous impurities.

Food laboratory safety programme - The strict observance of the precept that food and drink are not to be consumed in a chemical laboratory, nor stored in the same place as chemicals for laboratory use. This is especially important with regard to refrigerators, deep freezers and ice-making machines.

Free radical - A chemical intermediate having one unpaired electron that can initiate a chain reaction; such a reaction may be self-sustaining or liable to quenching by a scavenger.

Fuller's earth - *Adsorbent bentonite clay* (q.v.) used to control spills of chemicals (or, in the medical treatment of poisoning, to adsorb a chemical which has been ingested).

Full-face respirator with supplied air - A respirator that is connected by a hose to a pump. The safe use of the system depends upon the location of the pump (away from the hazard in question), the integrity of the hose and other factors.

Gavage - Administration of a material directly into the stomach of an experimental animal by intraoesophageal intubation.

Gauntlet-type rubber gloves - Rubber gloves that extend beyond the mid-arm up to the elbow.

Glove box - A sealed working area in the form of a box with gloves attached, designed to allow the handling of dangerous material, or protect the material from atmospheric moisture or oxygen. See also *Positive-pressure glove box* and *Negative-pressure glove box*.

Good Laboratory Practice - A desirable or mandatory system of laboratory working, incorporating basic safety instructions and appropriate experimental design features.

Grounded/grounding - The maintenance of the conducting framework of electrical apparatus at earth potential.

Hazard - The general term for any chemical, biological or physical agent which has the potential to cause injury. See also *Risk*.

Hazardous chemical - A substance that poses a danger to human health or the living environment if improperly handled. The hazard associated with each chemical is constant; *risk* (q.v.) is related to the concentration ("level") and duration of exposure.

Hazardous-waste coordinator - The person who has the responsibility for training laboratory or workplace personnel to deal with the most probable incidents, such as the plugging of a leak in a large container, clean-up of spillage, or transfer of a liquid from one container to another.

Hazardous-waste plan - A document which sets out all aspects of safe waste collection and disposal, and anticipates a variety of high risk situations (such as a leaking drum, flash fire, or splashing of a chemical) based on a knowledge of the type of waste. The document should specify an appropriate range of equipment, protective clothing, respiratory protection, and absorbent material to be available, so as to deal with such situations.

Heart (cardiac) resuscitation - Cardiac massage or stimulation. See also *Cardiopulmonary resuscitation*.

Hepatic insufficiency - Damage to the liver, or compromised liver function, as a result of medication, drug or alcohol abuse, chemical exposure, or disease due to a variety of other causes.

High-efficiency-particulate-air (HEPA) filter - A sub-micrometer sized filter used to retain particulate contaminants in air - often incorporated into microbiological safety cabinets, and into safety helmets worn by operators to protect against particle inhalation, etc.

High-efficiency scrubbers - The use of baffles to alter the flow of a gas or vapour to facilitate its purification by condensation, washing, etc.

Housekeeping - The maintenance of all laboratory facilities in a neat fashion, at hand and in good condition, with up-to-date records.

Housekeeping inspections - An assessment of the success in keeping the laboratory facilities in good condition, with records and other documentation up to date.

Hydrogen embrittlement - Structural changes leading to loss of strength that take place in carbon steel exposed to temperatures above 200 $^{\circ}$C or pressures of hydrogen greater than 21 MPa.

Hypersensitivity - See *Allergy*.

Ignition temperature - (Auto-ignition temperature) - The minimum temperature required to initiate or cause self-sustained combustion independent of the heat source.

Implosion - The catastrophic collapse of the walls of a closed container due to excessive external pressure in comparison with the internal pressure.

Incompatible chemicals - Two or more chemicals which can react with each other violently, with evolution of substantial heat, so as to produce flammable or toxic products.

Incompatible wastes - Two or more chemical wastes, the mixing of which can result in the slow or explosive release of toxic gases, vapours, or fumes. This may occur spontaneously at ambient temperature and pressure, or if the mixture is exposed to water, subjected to strong force, or heated under confined conditions.

Incineration - The controlled combustion of waste so as to prevent explosion or the unacceptable release of harmful products.

Indicating molecular sieve - A molecular sieve that indicates by a colour change when the capacity for water absorption in drying a gas or vapour has been exceeded.

Inflammation - Tissue response to injury that is characterized by local redness, swelling, pain and increase of temperature.

Instructions on reactions to possible emergencies - For example, a fire evacuation notice.

Inverse Joule-Thomson effect - The thermodynamic effect which occurs when the expansion of hydrogen or helium takes place at a temperature higher than the inversion temperature (-80 °C or -240 °C, respectively).

Irritant - A substance which has irritant properties or causes overt inflammation following immediate, prolonged or repeated contact with skin, mucous membrane, or other tissues.

Jumpsuit - A protective overall with elasticated arm and leg coverings; it may also include full feet covering and head protection. This kind of protective wear is commonly made of disposable paper or polymer-based material that is coated so as to be impervious to harmful liquids or other materials.

Kevlar® - A plastic material which is resistant to exposure to relatively high temperatures.

Laboratory - A building or part of a building used mainly by scientists, engineers or technicians for the investigation of the chemical, biological or physical properties of substances; for development of new or improved chemical processes, products or applications; for analysis, testing, or quality control; or for instruction and practice in natural science or in engineering.

Laboratory chief executive - Director or person in overall charge of a laboratory.

Laboratory manager - Person responsible for the planning and day-to-day management of the work of the laboratory.

Laboratory personnel - All persons whose workplace is the laboratory.

Laboratory supervisor - Person responsible for the day-to-day supervision of the work of a section of a laboratory.

Lab-pack - A 200-L open head steel drum that can be filled with small containers of chemicals packed in and separated by an inert absorbing medium, for the purpose of waste-disposal by landfill. A similar pack uses an outer fibre drum for the purpose of transportation to a waste-treatment facility or for incineration.

Lachrymator - A chemical that irritates the mucous membranes of the eye or nose and causes secretions (tears, or a running nose).

Landfill - The disposal of solid waste by burial in layers of earth in low ground or in disused mineworkings. A *secure landfill* (q.v.) is constructed so that *leachate* (q.v.) from the waste will not permeate strata so as to contaminate water supplies. See also *Sanitary landfill*.

Laundry bleach - Solution of sodium hypochlorite sold as a household bleach (e.g. Clorox®, which contains 5.25% w/v (0.75 mol/L) sodium hypochlorite). The trade names of such products and their nominal concentrations differ throughout the world.

LD_{50} *dose* - The dose of a chemical that is lethal to 50% of test animals when administered orally or by some other route in a single dose. The dose (LD_{50}) is expressed in grams or milligrams per kilogram of body weight.

Leachate - Soluble or suspended material that can percolate or be washed out of buried material in a landfill.

Lesion - A pathological disturbance resulting from an injury, an infection or a tumour.

Limits of flammability - These are expressed in terms of the *lower explosive limit* and *upper explosive limit*, each of which is defined separately.

Lower explosive (flammabile) limit - The concentration (percent by volume) of a defined vapour in air, below which a flame is not propagated when an ignition source is present. Below this concentration, the mixture is too lean to burn.

Malodorous substance - A substance that has an unpleasant smell.

Medical surveillance - The chemical monitoring of laboratory and other workplace personnel in the context of exposure to chemicals, or to assess the consequences of such exposure by clinical laboratory methods. See also *Biological monitoring.*

Methaemoglobinaemia - The presence of methaemoglobin in the blood due to exposure to a chemical that gives rise to oxidation of haemoglobin.

Molecular sieve - Highly adsorbent porous materials derived from heat-activated clay, diatomaceous earth etc. capable of trapping substances in the crystal lattices and often used to purify or dry gases or vapours.

Monitoring of airborne concentrations - A chemical technique used in occupational health surveillance to measure a variety of different toxic materials in workplace or ambient air. See *Personal air-sampler* and *Personal sampler.*

Negative-pressure glove box - A *glove box* (q.v.) that is maintained and used at a lower pressure than atmospheric to ensure that there is no loss of material inside it to the laboratory.

Neurotoxicant - A poison that selectively affects the nervous system.

Non-hazardous waste - Waste material of sufficiently low toxicity for safe disposal in a *sanitary landfill* (q.v.). All such materials have LD_{50} of 500 mg/kg or more.

Norit® - A commercially available form of activated charcoal.

Oedema (edema) - Accumulation of fluid in tissue as a result of injury, inflammation, allergy, etc.

Organophosphorus insecticides - A large group of derivatives of phosphoric acid that have in common their application to killing insects; they are potent inhibitors of *acetylcholinesterase* (q.v.), leading to the accumulation of acetylcholine, and this effect gives rise to associated neurotoxic changes in humans.

Orphan reaction mixture - An unlabelled material of unknown origin or generated by a former worker in the laboratory.

Personal air-sampler - A pump that draws ambient air through a filter or adsorbent chamber at a fixed pumping rate in order to estimate the likely concentration of an airborne chemical to which a worker has been exposed.

Personal safety habits - Individual practices consistent with good laboratory safety such as no eating, smoking, drinking, chewing gum or application of cosmetics in laboratory areas. One component of *Good Laboratory Practice* (q.v.).

Personal sampler (Individual monitor) - A compact, portable instrument for individual air-sampling in the respiration zone of a working person; used for the detection or measurement of harmful substances.

Pitot tube - An instrument that measures the effective static pressure of a stream of flowing fluid (e.g. air movement); also known as an impact tube.

Poison - A chemical substance or biological material which, taken into or formed within the organism, destroys life or impairs the health status.

Polymerization - Chemical bonding of monomers to form a polymer.

Positive-pressure glove box - A *glove box* (q.v.) in which the internal pressure is maintained higher than atmospheric, commonly using dry nitrogen, so as to exclude moisture or oxygen. See also *Glove box* and *Negative-pressure glove box*.

Protective apparel - Garments or accessories, such as shoes, laboratory coat, gloves or goggles that are intended to protect laboratory personnel from injury or chemical exposure.

Protective equipment - Any device, such as a respirator, safety shield, fume cabinet or manipulator, that is intended to protect laboratory personnel from injury or chemical exposure.

Pulmonary oedema - Collection of fluid in the lung. See also *Oedema*.

Pulmonary resuscitation - See *Artificial resuscitation*.

Pyrex® - A trademark for borosilicate glass.

Pyrophoric - The property of spontaneous ignition of a chemical in air.

Radioactive waste - Any material that is intrinsically radioactive, or contaminated with radioactive substance that needs to be disposed of in accordance with appropriate local, national or international regulations.

Regulatory requirements - Any legal or statutory legislation, which, in the context of this monograph, relates to laboratory safety, the safe use of chemicals, their storage, disposal, monitoring of laboratory personnel, etc.

Resinification - A process involving slow polymerization of low molecular weight substances, sometimes induced by exposure to air.

Return procedure - Part of a safety plan intended to ensure that personnel do not return to a laboratory area until an emergency situation is ended.

Rhinitis - Inflammation of nasal mucous membrane.

Risk - The likelihood of suffering a harmful effect resulting from exposure to a risk factor (usually some chemical, biological or physical agent). Risk may be expressed quantitatively as the probability of occurrence of an adverse effect, e.g. as the ratio between the number of individuals who would be expected to experience an adverse effect in a given time, and the total number of individuals exposed to the risk factor. The danger of the adverse consequences is related to the inherent hazard of the chemical in question and the level of exposure. The term "absolute risk" is sometimes used as a measure of risk expressed per unit dose (or exposure), or for a given dose (or exposure). Risk can be minimized by preventing or strictly limiting exposure to a chemical.

Risk assessment - The process of decision-making applied to a problem where there are several possible outcomes, and it is uncertain which event might actually occur. One component of risk assessment is risk estimation which includes:
(i) identification of the possible outcomes;
(ii) estimation of the magnitudes of the associated consequences of these outcomes; and
(iii) calculation of the probabilities of these outcomes.

Safety awareness - General appreciation by personnel of necessary safety requirements for the tasks in hand.

Safety coordinator/officer - Individual whose task it is to ensure safe day-to-day operation of a part or whole of the laboratory.

Safety and health organization - An established structure of responsible persons throughout a workplace, whereby safe working practices may be developed and promulgated.

Safe practices in laboratories - Accepted methods of working in a laboratory so as to comply with the relevant safety plan.

Safety training - Specific training given to all personnel to enable them to carry out their jobs safely.

Segregation of waste - Good practice in which incompatible wastes are transported and stored separately.

Salve - Soothing ointment for topical use.

Sanitary landfill - A landfill that is operated by a municipal or other responsible authority for the disposal of household and municipal waste that is not considered hazardous.

Sanitary sewer - A sewer that is connected directly to a water-treatment plant. See also *Storm-sewer system*.

Secondary containment - A secure system for containment of a hazardous chemical such as an outer tray, or the storing of a bottle in a specially designed resistant container.

Secure landfill - A *landfill* (q.v.) that is permitted by an appropriate local or national agency to receive and landfill waste that is defined by that agency as hazardous.

Self-contained respirators - Protective breathing equipment including a face mask connected to cylinders of compressed air that are carried on the back or chest.

Shock-sensitive - A chemical that is sensitive to impact, or where the energy imparted to it by friction is likely to result in an explosion.

Short-term exposure limit - The time-weighted average airborne concentration of a chemical to which a worker may be exposed for periods of up to 15 minutes, with no more than 4 such excursions per day and at least 60 minutes between them.

Shut-down procedure - A set of guidelines which should be known to all personnel for shutting down an operation in an emergency. See also *Start-up procedure*.

Smoke source or tube - A device used to generate a non-toxic smoke with particle size 0.01-5 µm that allows the flow of air to be visualized in a fume hood or similar apparatus.

Spark-free appliance - Electrical apparatus specially designed to avoid the production of sparks in normal use. Most domestic appliances, such as mixers, driers, blenders and refrigerators are not constructed to spark-free standards and should not be used in laboratories where flammable materials may be present.

Spasm of the larynx - Allergic or other contraction of laryngeal muscles that makes breathing difficult.

Spill-control centre - A collection of all the materials that are necessary to safely contain and control a spilled chemical and prepare it for safe disposal. Spillage kits, with instructions, appropriate absorbents and reactants, together with protective equipment, should be located in strategic positions around a work area.

Spontaneous ignition or combustion - A fire which occurs when a substance reaches its ignition temperature without the local application of an external heat source.

Start-up procedure - A series of steps that must be followed in sequence for the safe operation of a system, or a piece of equipment, or for a special operation. The procedure should be displayed prominently and reviewed regularly. See also *Shutdown procedure.*

Static discharge - Electrical discharge or spark due to a static electrical charge accumulating on an object; static-generated sparking should be avoided by using a *Grounding (earthing)* strap.

Stockroom personnel - Individuals with general responsibility for ordering, transporting, and storing chemicals; for keeping relevant records up to date; for issuing supplies of chemicals; and sometimes, also, for organizing the safe disposal of unwanted or waste chemicals.

Storm-sewer system - Discharge of storm or waste water or sewage directly into a river, lake or sea without any treatment.

Supplementary-air hood - Mechanical device that directs a blanket of air vertically between the operator and the sash of a hood.

Supplied-air respirator - A life-support system that brings fresh air through a length of hose to the facepiece of a respirator at a pressure high enough inside the mask to be slightly greater than atmospheric.

Syrup of ipecac (ipecacuanha) - Extract of a South American shrub sometimes used to induce vomiting when an individual has swallowed a non-corrosive or petroleum-free poison.

Systemic toxic effects - Toxicity to the body as a whole. Systemic effects require the absorption and distribution of a toxic substance to sites distant from the entry point. Most chemicals that produce systemic toxicity do not cause a similar degree of toxicity in all organs, but usually demonstrate major toxicity to one or two organs, which are referred to as the target organs for that chemical.

Tachycardia - Abnormally rapid heart rate (over 100 beats per minute).

Teflon ® - A trademark for a polymer made from tetrafluoroethene that is highly resistant to attack by most solvents and other chemicals.

Teratogen - A chemical that has the potential to cause deformities in the fetus.

Thermally-broken-down hazardous waste - Non-hazardous chemicals produced by heat treatment of waste.

Thermodynamic instability - Property of a chemical intermediate that causes it to break down, often with the release of energy leading to heat or explosion.

Threshold limit values (TLVs) - Recommended guidelines for occupational exposure to airborne contaminants published by the American Conference of Governmental Industrial Hygienists; a TLV (in mg m^{-3}) for an 8-hour workday and a 40-hour work week to which nearly all workers may be repeatedly exposed, day after day, without expected adverse effects.

Toxicity - The capacity of a chemical or biological material to cause injury to a living organism. A highly toxic substance will cause damage to an organism if administered in very small amounts; a substance of relatively low toxicity will not produce an effect unless the amount is large. However, toxicity cannot be defined in quantitative terms without reference to: the quantity of substance administered or absorbed; the way in which this quantity is administered (e.g. inhalation, ingestion, injection); its distribution in time (e.g. single or repeated doses); the type and severity of the injury in question; and the time needed to produce the injury.

Toxin - A poisonous substance produced by a biological organism such as a microbe, plant or animal).

Toxicology - The understanding of the effects of toxins and poisons

Treatment site - A location or plant where waste is processed to reduce its toxicity and/or the volume it occupies for disposal.

Tumour - A malignant mass of cells or tissue.

Tygon® - A trademark for a chemically resistant plastic, often used for tubing, e.g. on an automatic analyser.

Unattended operation - The practice of leaving equipment to run outside normal working hours; such equipment should be fitted with suitable failsafe devices. See also *Failsafe design*.

Unguent - Ointment (salve) for topical application.

Unsafe condition - Any omission of a safe practice or safe working condition that increases hazard or risk.

Upper explosive limit (upper flammability limit) - The maximum concentration (percent by volume) of the vapour in air above which a flame is not propagated. See *Limits of flammability* and *Lower explosive limit (lower flammability limit)*.

Vacuum-distillation - Distillation under reduced pressure to achieve a lower boiling point and reduce thermal degradation.

Vaporization - Volatilization of a liquid at temperatures below the boiling point.

Vapour pressure - Pressure of a vapour in contact with its liquid or solid phase at a defined temperature.

Vehicle drivers and disposal operators - Individuals employed in transportation and disposal of chemicals who should know of the hazards of the wastes they are handling.

Velometer - See *Anemometer*.

Ventilation system - A means of supplying fresh air to and disposing of used air from the laboratory environment.

Vermiculite - Hydrous magnesium-aluminium-iron silicate in an expanded form derived from mica; used as a heat insulator and inert packing material with high absorptive capacity for liquids.

Volatile substance - A substance with a high tendency to become vapour at ambient temperature and pressure.

Waste - Any solid, liquid or gas that is a by-product, residue or unwanted, excess from a chemical reaction; waste may require special disposal procedures, especially if it is hazardous.

Waste-management system - A clearly defined and documented system of operating procedures which together provide for safe handling and disposal of laboratory waste, regardless of its physical form. The system must comply with all relevant regulatory requirements.

Waste-water effluent - Waste water arising from laboratory operations designated for disposal to a sewage system.

Waste-water treatment plant - An engineering installation specially designed to treat waste water by chemical or other means before disposal as non-toxic waste to a sewage system.

Warning sign - An advisory or statutory sign that identifies a chemical, physical or other hazard.

Water-reactive chemical - A compound or element that reacts rapidly or violently with water.

Xenobiotic - A substance that is foreign to the target organism or organ - it is an exogenous as opposed to endogenous chemical. Most toxic substances and drugs are properly described as xenobiotics.

Bibliography

Alcock, P.A. (1982). Safety Inspections — the Detection of Hazards at Work. London: H.K. Lewis & Co.

Arena, J.M. & Drew, R.H. (eds.) (1986). General considerations of poisoning. In Poisoning — Toxicology, Symptoms, Treatments, 5th edn., Chapter 1, pp. 3—173. Springfield: Charles C. Thomas.

Baselt, R.C. (1982). Disposition of Toxic Drugs and Chemicals in Man. Davis: Biomedical Publications.

deBethizy, J.D. & Hayes, J.R. (1989). Metabolism: a determinant of toxicity. In Principles and Methods of Toxicology (ed., A.W. Hayes), 2nd edn., Chapter 2, pp. 29—66. New York: Raven Press.

Bretherick, L. (ed.) (1986). Hazards in the Chemical Laboratory, 4th edn. London: Royal Society of Chemistry.

Bretherick, L. (1990). Handbook of Reactive Chemical Hazards, 4th edn. London: Butterworths.

Burns, D.T. Townsend, A., Carter, A.H. (1981) Inorganic Reaction Chemistry, Vol. 2, Ellis Horwood, New York.

Clayton, G.D. & Clayton, F.E. (eds.) (1981). Patty's Industrial Hygiene and Toxicology, 3rd edn., Volumes 2A, 2B & 2C, *Toxicology*. New York: John Wiley & Sons.

Committee on Hazardous Biological Substances in the Laboratory (National Academy of Sciences) (1989). Safe disposal of infectious laboratory waste. In Biosafety in the Laboratory — Prudent Practices for the Handling and Disposal of Infectious Materials, Chapter 4, pp. 34—45. Washington D.C.: National Academy Press.

Croner Publications (1988). Croner's Hazardous Waste Disposal Guide. New Malden: Croner Publications.

Ellenhorn, M.J. & Barceloux, D.G. (1988). Chemical products. In Medical Toxicology — Diagnosis and Treatment of Human Poisoning, Part IV, pp. 779—1108. New York: Elsevier Science Publishing Company.

Erdey, L. (1965) Gravimetric Analysis, Part II, Pergamon Press, New York;

Fawcett, H.H. (1984). Hazardous and Toxic Materials—Safe Handling and Disposal. New York: John Wiley & Sons.

Forsberg, K. & Keith, L.H. (1989). Chemical Protective Clothing Perfor-

mance Index Book. Chichester: John Wiley & Sons.
Freeman, N.T. & Whitehead, J. (1982). Introduction to Safety in the Chemical Laboratory. London: Academic Press.
Fuscaldo, A., Erlick, B.J. & Hindman, B. (eds.) (1980). Laboratory Safety —Theory and Practice. New York: Academic Press.
Grisham, J.W. (ed.) (1986). Health Aspects of the Disposal of Waste Chemicals. New York: Pergamon Press.
Haley, T.J. & Berndt, W.O. (eds.) (1987). Toxicology. Washington: Hemisphere Publishing Corporation.
Hawkins, M.D. (1988). Safety and Laboratory Practice, 3rd edn. London: Cassell Publishers.
Houston, A. (ed.) (1986). Dangerous Chemicals —Emergency First Aid Guide. New Malden: Wolters Samson (United Kingdom).
Hughes, D. (1989). Discharging to Atmosphere from Laboratory—Scale Processes, HHSC Handbook No. 4. Leeds: H and H Scientific Consultants.
International Agency for Research on Cancer (1982/83/84/85). Laboratory Decontamination and Destruction of Carcinogens in Laboratory Wastes. IARC Scientific Publications: No. 43, Some *N*—Nitrosamines; No. 49, Some polycyclic aromatic hydrocarbons; No. 54, Some hydrazines; No. 55, Some *N*-Nitrosamides; No. 61, Some haloethers; No. 73, Some antineoplastic agents. Lyon: International Agency for Research on Cancer.
Jackson, H.L., McCormack, W.B., Rondesvedt, C.S., Smeltz, K.C. and Viele, I.E. (1974) Safety in the Chemical Laboratory, Vol. 3 (ed. N.V. Steere), American Chemical Society, Easton, PA.
Klaasen, C.D., Amdur, M.O. & Doull, J. (eds.) (1986). General principles of toxicology. In Casarett and Doull's Toxicolgy—The Basic Science of Poisons, 3rd edn., Unit I, pp. 1—220. New York: Macmillan Publishing Company.
Leinster, P. & Evans, M.J. (1986). The application, generation, and validation of human exposure data. In Toxic Hazard Assessment of Chemicals, Chapter 8, pp. 72—83. London: Royal Society of Chemistry.
Lenga, R.E. (ed.) (1988). The Sigma—Aldrich Library of Chemical Safety Data, 2nd edn. (2 Volumes). Milwaukee: Sigma—Aldrich Corporation.
Lunn, G. & Sansone, E.G. (1990). Destruction of Hazardous Chemicals in the Laboratory. Chichester: John Wiley & Sons.
Menendez—Botet, C.J. & St. Germain, J.M. (eds.) (1990). Hazardous waste —Facts and Fallacies. Washington D.C.: American Association for Clinical Chemistry, AACC Press.
Pal, S.M. (ed.) (1990). Handbook of Laboratory Health and Safety Measures, 2nd edn. Dordrecht: Kluwer Academic Publishers.

Pipetone, D.A. (ed.) (1991). Safe Storage of Laboratory Chemicals, 2nd edn. New York: John Wiley & Sons.

Procter, N.H., Hughes, J.P. & Fischman, M.L. (eds.) (1988). Chemical Hazards of the Workplace, 2nd edn. Philadelphia: J.B. Lippincott Company.

Royal Society of Chemistry (1988/89/90). Chemical Safety Data Sheets — Solvents (Vol. 1); Main Group Metals and their Compounds (Vol. 2); Corrosives and Irritants (Vol. 3). Cambridge: Royal Society of Chemistry.

Royal Society of Chemistry (1989). Safe Practices in Chemical Laboratories. London: Royal Society of Chemistry.

Royal Society of Chemistry (1990). Guidance on Laboratory Fume Cupboards. London: Royal Society of Chemistry.

Sax, N.I. & Lewis, R.J., Sr. (eds.) (1986). Rapid Guide to Hazardous Chemicals in the Workplace. New York: Van Nostrand Reinhold Company.

Sax, N.I. & Lewis, R.J., Sr. (1989). Dangerous Properties of Industrial Materials, 7th edn. (3 Volumes). New York: Van Nostrand Reinhold Company.

Scott, R.M. (1989). Chemical Hazards in the Workplace. Chelsea: Lewis Publishers.

Sinclair, W. (1988). Radiation. Chemical hazards. Biological hazards. In The Hazards of Hospital Work, Chapters 4, 5 & 6, pp. 62—164. Sydney: Allen & Unwin.

Sittig, M. (1981). Handbook of Toxic and Hazardous Chemicals. Park Ridge: Noyes Publications.

Steere, N.V. (ed.) (1990). CRC Handbook of Laboratory Safety, 3rd edn. Boca Raton: CRC Press.

Stricoff, R.S. & Walters, D.B. (1990). Laboratory Health and Safety Handbook — a Guide for the Preparation of a Chemical Hygiene Plan. Chichester: John Wiley & Sons.

Task Force on the Resource Conservation and Recovery Act (American Chemical Society) (1990). The Waste Management Manual for Laboratory Personnel. Washington, D.C.: American Chemical Society.

Warren, P.J. (ed.) (1985). Dangerous Chemicals — Emergency Spillage Guide. New Malden: Wolters Samson (United Kingdom).

Young, J. (ed.) (1988). Improving Safety in the Laboratory — A Practical Guide. Chichester: John Wiley & Sons.

Chemical Index

Absorbent mixture	44	metals	249
Acetaldehyde	24, 114	oxides	249
diethyl acetal	244	peroxides	249
Acetals	117, 244	Alkanedioic acids	252
Acetic acid	24	Alkanediols	252
Acetone	24, 114, 121, 170, 247	Alkanes	244
Acetonitrile	253	Alkanoic acids	252
2-Acetylaminofluorene	102	Alkanols	252
Acetyl peroxide	175	Alkoxyalkanols	252
Acetylcholinesterase	97	Alkyl	
Acetylene	83, 114, 126, 137, 146, 194, 247, 249	alkanesulfonates	164
		chlorates(VII)	200, 246
Acetylenic compounds	126, 246	halides	164
Acetylides	117, 121, 146	nitrates	246
Acid		nitrites	246
anhydrides	168	phosphates	169
chlorides	79	sulfates	169
halides	165, 168, 190	Alkylacetylenes	244
Acids	249	Alkylaluminiums	192
Acrolein	117, 170	Alkylarenes	244
Acrylates	244	Alkylating agents	164, 167, 169
Acrylic acid	245	Alkyllithiums	192, 218
Acrylonitrile	24, 137, 170, 224, 245	Allergens	92
Acute hazardous wastes	206	Alloys	189
Acyl		Allyl	
halides	247	chloride	164, 224
nitrates	246	halides	164
nitrites	246	Aluminium	178, 189
Acylating agents	169	alkyls	82, 241
Aflatoxin(s)	162, 236	chloride	126
B1	102	hydride	179
Alcohols	165, 218, 238, 252	nickel alloy	192
Aldehydes	170, 218, 244, 252	oxide	177
Aliphatic		Amalgams	158
amines	172, 252	Amides	169, 179, 244
diamines	252	Amine metal oxosalts	246
Alkali		Amines	172, 182, 223, 246, 247, 252
carbides	249		
hydrides	249	Amino acids	234
hydroxides	249	Aminoalkanoic acids	253
metal amides	241	4-Aminobiphenyl	172
metal hydrides	241	Ammonia	93, 114, 126, 130, 134, 137, 182, 246, 247, 249
metals	30, 79, 188, 241, 244, 245, 249		
		Ammonium	
oxides	249	chlorate(VII)	199
peroxides	249	hydroxide	24
Alkaline earth		nitrate(V)	184
carbides	249	t-Amyl alcohol	252
hydrides	249	Amyl nitrite	101
hydroxides	249	Anhydrides	241

273

274 Chemical Safety Matters

Anhydrous		solid	18, 126, 142, 145
ammonia	147, 249	Carbon disulfide	24, 114, 126,
metal halides	241		168, 209, 247
Aniline	24	Carbon monoxide	115, 148
Anions	177	Carbon tetrachloride	24, 87, 127,
dangerous	182		164, 224, 247
Antimony	178	Carbonates	179, 234
Antineoplastic agents	162	Carboxamides	169
Arenediazonium salts	176	Carboxylic acids	168, 252
Arenesulfonates	164	Cations	177
Aromatic amines	162, 172	dangerous	182
Arsenate	179	metalloids	177
Arsenic	178, 182	metals	177
Arsenic acid	168	Cellosolves	167
Aryllithiums	192, 218	Cerium	178
Arylsodiums	192	Cesium	178, 244
Asbestos	23	Chlorate	126, 207
Atropine	97	Chlorate(I)	179, 182, 246
Azides	117, 121, 177, 179,	Chlorate(V)	112, 179, 182, 246
	182, 218, 246, 249	Chlorate(VI)	248
Aziridines	164	Chlorate(VII)	128, 177, 179, 246, 247
Barium	178	Chloric(VII) acid	25, 93, 128, 184, 199
Benzaldehyde	24	Chloride	178, 179, 234
1,2-Benzanthracene	163	Chlorinated	
Benzene	24, 89, 114, 224	dibenzodioxins	164
Benzoanthracene	102	organic compounds	223
3,4-Benzpyrene, see benzo[a]pyrene		Chlorine	24, 126, 134, 148, 238, 247
Benzopyrene, see benzo[a]pyrene		Chloroacetic acid	165
Benzo[a]pyrene	102, 104, 163	Chloroalkanedioic acid	253
Benzoyl peroxide	126, 175	Chlorobenzene	164, 224
Benzyl halides	164	2-Chloroethanol	165, 252
Benzylamine	252	Chloroform	24, 121, 127, 164, 224, 247
Beryllium	177, 178	Chloromethyl ethers	164, 165
Bis(chloromethyl) ether	89, 91, 102, 103	Chloroolefins	244
Bismuth	178	Chlorophenols	165
Borates	179, 234	Chloroprene	245
Borohydrides	179	Chromate	92, 119
Boron(III) fluoride	134, 148, 186, 190	Chromate(VI)	179, 182, 183,
Boronic acids	168		196, 246, 248, 249
Bromate(V)	179, 182	Chromic(V) acid	93
Bromide	179	Chromic(VI) acid	24, 93, 249
Bromine	24, 177	Chromium	182
Bromobenzene	164	Chromium(III)	178
Butadiene	137, 245	Chromium(VI)	
1,4-Butanediamine	253	oxide-pyridine complex	126
t-Butyl chloride	164	trioxide	248
t-Butyl hydroperoxide	175	Citric acid	234
Butyraldehyde	24	Cobalt	127, 189
Butyric acids	253	Cobalt(II)	178
Cadmium	178	Copper	177, 178, 247
Calcium	178	phosphide	191
carbide	194, 241	Copper(I) cyanide	187
chlorate	185	Crown ethers	167
chlorate(I)	247	Cumene	244
fluoride	186, 190	Cyanate	100, 179
hydride	185	Cyanide	40, 99, 179, 182, 249
molybdate	178	Cyanogen	99
oxide	93	bromide	187
sulfate	177	halides	99
tungstate	178	Cycloalkanes	244
Carbides	79	Cyclohexane	24, 114
Carbitols	167	Cyclohexene	117, 163, 244
Carbon	248	Cyclopentene	244
Carbon dioxide	18, 130, 134	Decahydronaphthalene	244

Chemical index

Decalin	244
Dehydrohalogenation	165
Diacetylene	244
Diacyl peroxides	117, 176
Dialkyl peroxides	117, 118, 119, 176
Dialkyl sulfates	164
Dialkylcadmium	193
Dialkylnitrosamine	173
Dialkylzincs	192
Diazo compounds	121, 246
Diazoalkanes	174
Diazomethane	92, 121, 126
Diazonium salts	199, 246
1,2-Dibromoethane	164
Di-t-butyl peroxide	175
Dibutyl phthalate	24
Dichloroacetylene	129
1,2-Dichlorobenzene	164
2,7-Dichlorodibenzo-p-dioxin	224
1,2-Dichloroethane	24, 224
1,2-Dichloropropane	24
Dichlorodiphenyltrichloroethane	224
Dicyclopentadiene	244
Dicyclopentadienyl iron complex	192
Dienes	244
Diethyl ether	24, 114, 117, 167, 185, 217, 244, 247
Diethylene glycol dimethyl	245
Difluoroamino compounds	246
Diglyme	117, 167
Diisobutyryl peroxide	175
Diisopropyl ether	175, 199, 244
Diisopropyl fluorophosphate	96
Diisopropyl peroxydicarbonate	175
Dimethyl sulfoxide	24, 126, 247
Dimethylamine	253
Dimethylcarbamoyl chloride	102
1,1-Dimethylhydrazine	192
Dimethylmercury	102, 106
N,N-diethylnitrosamine	107
N,N-dimethylacetamide	169
N,N-dimethylformamide	169
N,N-dimethylnitrosamine	192
Dinitroacetonitrile	247
p-Dioxane	117, 167, 253, 245
Dioxolane	253, 245
Divinylacetylene	244
Dry ice, see carbon dioxide, solid	
Dusts	115
Dyes	176
Epoxides	164, 167
Esters	169, 253
Ethanol	79, 87, 91, 114
Ethers	117, 167, 244, 253
Ethyl acetate	24, 185
Ethyl acrylate	245
Ethyl bromide	164
Ethylene glycol	25
dimethyl ether	244
ether acetate	244
monoether	244
Ethylene oxide	127, 238
Ethylmercuric chloride	106
Ferricyanides	179
Ferrocene	192
Ferrocyanides	187
Flowers of sulfur	115
Fluorides	179, 182, 234
Fluorine	25, 136, 149
Fluoroacetic acid	165
Fluoroolefins	244
Formaldehyde	25, 87, 92, 170, 172, 217, 238
Formamide	91
Formic acid	25
Fulminates	246
Fungicides	237
Furan	244
Gallium	178, 182
Germanium	178
Glycerol	252
Glyme	117, 167
Gold	177, 178
Grignard reagents	192, 218, 241, 245
Guanidine	246
Hafnium	178
Halides of non-metals	241
Haloacetylenes	246
Haloamines	121
Haloethers	162
N-Halogen compounds	246
Halogen azides	246
Halogenated	
hydrocarbons	164, 188, 218
wastes	209
Halogenating agents	248, 249
Halogens	248, 249
Heavy metal(s)	102, 177, 249
chlorates(VII)	199
n-Heptane	114
2,2′,4,4′,5,5′-Hexachlorobiphenyl	224
Hexachlorobenzene	224
Hexamethylphosphoramide	102, 169
n-Hexane	25, 114
Hydrazines	126, 162, 192
substituted	192
Hydrazoic acid	187, 188
Hydride	177, 179, 182
Hydrobromic acid	25
Hydrocarbons	163, 218
Hydrochloric acid	25, 89
Hydrofluoric acid	25, 93, 97
Hydrogen	114, 132, 150
sulfite, see sulfate(IV)	
bromide	151
chloride	134, 151
cyanide	91, 96, 99, 186, 196
embrittlement	143
fluoride	96, 186
peroxide	25, 117, 127, 184, 246, 248
sulfide	91, 114, 134, 152, 196
Hydrogenation	137
catalysts	127, 189
Hydroperoxides	117, 118, 119, 179, 247
Hydrosulfide	179
Hydroxides	179
of cations	251

Hydroxyalkanoic acids	252	azides	187
β-Hydroxypropionic acid	169	carbonyls	192, 193, 245
Hydroxylamine	246	hydrides	29, 79, 82, 184, 245, 248
Indium	178	nitrates	184
Inorganic		phthalocyanines	192
acid halides	241	powders	245
acids	44	Methacrylates	169, 244
bases	44	Methanol	34, 114
cyanides	170, 186	2-Methoxyethanol	25
fluorides	186	Methyl acrylate	96, 137
hydroperoxides	182	3-Methylcholanthrene	102
peroxides	182, 207	N-Methyl-N'-nitrosoguanidine	174
sulfides	185	N-Methylpyridinium-2-aldoxime	97
wastes	177	Methyl chloride	224
Insecticides	237	Methyl ethers	127
Insoluble sulfides	180	Methyl ethyl ketone	114, 170
Iodate(V)	179, 182, 246	Methyl iodide	164
Iodate(VII)	182	Methyl isobutyl ketone	244
Iodide	179	Methyl ketones	172
Iodine	126, 238	Methyl methacrylate	245
Iridium	178	Methyl n-butyl ketone	172
Iron	178, 189	Methyl vinyl ketone	172
Iron(II) sulfate	184	Methylacetylene	244
Isocyanates	92	Methylcyclopentane	244
Isopropyl		Methylene chloride	25, 127, 164
acetate	253	Methylene iodide	164
alcohol	114	Methylhydrazine	192
Ketones	170, 218, 253	Molybdate(VI)	182
Lactams	244	Molybdenum	178
Lactic acid	234	Monoethanolamine	25
Lanthanides	178	Morpholine	25
Lead	177, 178	2-Naphthylamine	94, 102, 172
azide	187	N-Nitro compounds	246
compounds	91	N-Nitromethylamine	246
Light metal	177	2-Nitronaphthalene	102
Liquid air	18, 115	N-Nitrosamides	162, 174
Liquid nitrogen	18, 115, 127	N-Nitrosamines	102, 162
Lithium	29, 178, 189	N-Nitroso compounds	173
alkyls	241	N-Nitrosodiethylamine	107
aluminium	127	N-Nitrosodimethylamine	104, 107
aluminium hydride	128, 184	Neutralizing agent	43, 44
Magnesium	29, 79, 178, 189	Nickel	92, 178, 189
powder	115	carbonyl	102
Manganates(VII)	112, 127, 179, 182, 183, 196, 207, 246, 248, 249	Niobium	178
		Nitrates	207
Manganese	182, 189	Nitrate(III)	179, 248, 249
Manganese(II)	178	Nitrate(V)	179, 246, 248, 249
Mercaptans	218	Nitric amide	246
Mercury	89, 105, 137, 157, 177, 178, 182, 249	Nitric(V) acid	25, 93, 112, 184, 247, 249
		Nitriles	169, 170, 223, 253
amalgams	249	Nitro compounds	121, 172, 218
compounds	104	Nitrocellulose	246
poisoning	106	Nitrogen	130
salts	106	chlorides	126
Mercury(II) sulfide	158	triiodide	126
Metal(s)	248	Nitrogen(IV) oxide	91
acetylacetonates	192	Nitroglycerine	246
acetylides	126, 146	Nitroguanidine	246
alkanecarboxylates	192	Nitroso compounds	121
alkoxides	192, 244	Nitrosomethylurea	174
alkyls	29, 30, 241, 245	Nitrourea	246
amides	179, 182, 188, 244	Noble metals	159
aryls	245	Non-metal	
		alkyls	193, 245

Chemical index

aryls	193
hydrides	190, 241
Olefins	117, 244
Organic	
acid halides	241
acids	168
acyl halides	249
anhydrides	249
compounds	248
halogen compounds	249
hydrocarbons	223
hydroperoxides	184, 199
nitro compounds	249
peroxides	117, 174, 176, 199
Organo-	
inorganic chemicals	192
mercurials	91, 193
metallics	30, 161, 192, 218, 244
phosphorus insecticides	96
sulfur compounds	168
Osmium	178, 182
tetroxide	178
Oxalic acid	249
Oxides	179, 234
Oxygen	83, 115, 127, 136, 153
Ozone	127, 238
Ozonide	121
Palladium	127, 159, 178, 189
Pentane	114
Perchloric(VII) acid	
see chloric(VII) acid	
Peroxide(s)	112, 117, 128, 159, 163, 167, 174, 179, 194, 218, 247, 248
content	118
forming chemicals	203, 205, 242
reagents	175
Peroxidizable compound	243
Phenols	25, 90, 92, 165, 218, 238, 247
Phenylhydrazine	192
Phosgene	153
Phosphate	179, 234
Phosphate(III)	182
Phosphides	79
Phosphine	128, 191
Phosphonates	169
Phosphonic acids	168
Phosphoric(I) acid	128
Phosphoric(III) acid	191
Phosphoric(V) acid	191
Phosphorus	112, 128, 191, 248, 249
white	190, 245
red	191
Phosphorus(III) chloride	128
Phosphorus(V) oxide	93, 127, 191, 241, 250
Picric acid	173, 199, 247
Pigments	176
Platinum	127, 159, 178, 189
oxide	127
Polyacetylenes	246
Polycarbonate	39
Polychlorinated biphenyls	164
Polycyclic aromatic	
hydrocarbons	104, 162
Polyhalogenated compounds	164, 225
Polymethyl methacrylate	39
Polymerization	99, 137, 186, 245
catalysts	82
Polymers	218
Polynitro	
alkyl compounds	247
aromatic compounds	199, 225, 247
compounds	173
hydrocarbons	247
Polyol nitrates	246
Polyvinyl chloride	39
Potassium	29, 82, 112, 178, 189, 244
amide	188, 200, 244
cyanide	100, 187
fluoride	186
hydroxide	93
metal	128, 244
sulfide	185
β-Propiolactone	103, 169
Propane	130
1,3-Propane sultone	102
Propionitrile	253
Pyridine	172, 252
Pyridinium chlorate(VII)	200
Pyrophoric	
chemicals	203, 245
metals	112
Quaternary ammonium salts	238
Raney nickel	127
Rhenium	178
Rhodium	159, 178
Rubidium	178, 244
Ruthenium	159, 178
Scandium	178
Selenate	179
Selenide	179
Selenium	178, 182
Silicon	248
Silver	158, 178, 249
fulminate	246
nitrate(V)	247
oxide	247
Sodium	29, 82, 112, 177, 178
amide (sodamide)	188, 200, 244
azide	187, 200, 247, 249
borohydride	184
chlorate	25, 185, 247
cyanide	100, 187
fluoride	186
hydride	185
hydroxide	25, 93
metal	129
sulfide	180, 185
thiosulfate	101
Solvents,	
see individual compounds	
Starch	234
Stibonic acid	168
Strontium	178
Strychnine	96
Styrene	114, 245
Sugar alcohols	234, 252
Sugars	234, 252

Sulfate(IV)	179, 234	Titanium(IV) chloride	57
Sulfate(VII)	112, 119, 179, 182, 248	Toluene	114, 185, 224
Sulfide	168, 178, 179, 182, 249	Trialkylaluminiums	192
inorganic	250	Triaminophenol	173
insoluble	180	Trichloroethylene	25, 129
Sulfite		1,1,1-Trichloroethane	164
see sulfate(IV)		1,1,2-Trichloro-1,2,2-trifluoroethane	164
Sulfones	168	Tricresyl phosphate	25
Sulfonic		Triethylamine	172, 224
acid	168, 253	2,4,6-Trimethylpyrylium	
acid esters	169	chlorate(VII)	200
Sulfonyl halides	168	2,4,6-Trinitrophenol	173
Sulfoxides	168	Trinitrotoluene	25, 172
Sulfur	248	Tritium	55
flowers of	115	Tungsten	178
heterocycles	168	Uranium	189
Sulfur(I) chloride	190	Ureas	244
Sulfur(IV) oxide	134	Valeric acid	253
Sulfuric acid	25, 92, 93, 129, 249	Vanadium	178
Tantalum	177, 178	Vinyl	
Tellurium	178	acetate	224, 245
Tetrachloroethylene	164	chloride	94, 224, 245
Tetrafluoroethylene	245	esters	244
Tetrahydrofuran	117, 127, 167, 185, 244, 253	ethers	244, 245
		halides	244
Tetrahydronaphthalene	117, 245	Vinylacetylene	244, 245
Tetralin	245	Vinylidene chloride	244
Tetranitromethane	247	Vinylpyridine	245
Thalidomide	91	Volatile acids	60
Thallium	177, 178, 182	p-Xylene	114
Thioacetamide	169	Yttrium	178
Thioamides	168	Zinc	178, 189
Thiocyanate	179	dust	115
Thiols	167	Zirconium	178, 189
Tin	178, 189		
Titanium	178, 189		

General Index

Abdominal pain	46	Barricades	122, 124, 127, 137
Absorbent		Bench shields	90
materials	231	Beverages	66
mixtures	44	Biological	
Accident	40	hazards	18
reporting	19, 42	pretreatment	218
Acetylenic compounds	126	wastes	236
Acid		Blankets	39
anhydrides	168	Bleeding	46
halides	165, 168, 190	Boiling points	112
Acrolein	170	Burning rates	111
Acrylonitrile	170	Burns	93, 97, 148, 149, 184, 186
Activated carbon	147, 165	Cadmium	178
Acute		Calcium	178
hazardous wastes	206	carbide	194, 241
toxicity	90, 91, 94, 96	chlorate(I)	185, 247
Acylating agents	169	fluoride	186, 190
Aerosol(s)	89, 95, 108	hydride	185
generators	57	molybdate	178
Air		oxide	93
baths	68, 73	sulfate	177
flow paths	51	tungstate	178
respirator	153	Caps	67
Aliphatic amines	172	Carbon	248
Alkali metals	188	Carbon dioxide	18, 130, 134
Alkylating agents	164, 167, 169	solid	18, 126, 142, 145
Allergens	92, 165, 169	Carbon disulfide	209
Alloys	189	Carcinogens	23, 26, 55, 82, 86, 89,
Amalgamation	137		94, 102, 104, 107, 162, 163,
Amalgams	158		164, 169, 172, 192, 237, 238
Amides	169	Cardiac resuscitation	45
Ampoules	96	Cardiopulmonary resuscitation	45
Anhydrides	241	Cartridge respirators	34
Anhydrous		Cation	177
ammonia	147, 249	of metalloids	177
metal halides	241	of metals	177
Animal		Charcoal	
carcasses	239	activated	147, 165
studies	108	Chemical(s)	
Anions	177	compatibility	221
Antineoplastic agents	162	contamination	48
Arenediazonium salts	176	of biological waste	237
Armoured hoods	124	decontamination	103
Aromatic amines	172	destruction	162
Asbestos	23	incompatibilities	249, 250
Ash	221	ingestion	48, 78, 89, 105, 186
Atropine	97	inhalation	78, 89, 105, 186
Auto-ignition temperature	112	interactions	88
Auto-transformers	64, 68, 71	procurement of	77
Autoclaves	137, 141, 238	pyrophoric	189, 191, 193,
Back injuries	46		203, 205, 245

Chemical Safety Matters

Chemicals (Contd)
  reactive 79, 115
  recycling of 157
  re-use of 157
  shock-sensitive 121, 246
  spills 43
  skin contact 48
  toxic 52, 67, 79, 82, 86, 87, 88, 91, 92, 152, 161, 178,
Chemicals as fuel 161
Chest pain 46
Chlorinated
  compounds 223
Chlorinated
  dibenzodioxins 164
  organic compounds 223
Chronic toxicity 91, 101, 104
Classification of laboratory waste 206
Closed-systems 18
Clothing footwear 24
Cold
  rooms 15, 64, 97
  traps 18, 145
Combustible materials 111
Compatible chemicals 208
Compressed
  gas cylinders 130
  gases 78, 79, 82, 84, 86, 130, 146
Contact lenses 21, 36
Contact with skin 89, 90, 105
Containment 213
Controlled substances 79
Convulsion 46
Copper tubing 136
Corrosive
  chemicals 92
  gases 151
  liquids 94
Corrosivity 207
Cosmetics 95
Crown ethers 167
Cryogenic
  hazards 18
  liquids 142
Cumulative poisons 89
Cylinder valve outlet 133
Cytotoxic drugs 91
Dangerous anions 182
Dangerous cations 182
Dehydrating agents 92
Dehydrohalogenation 165
Desiccators 145
Destruction
  efficiency 224
  of hazardous chemicals 162
  procedures 162
Detoxification of hazardous wastes 233
Dewar flasks 17, 145
Diacyl peroxides 176
Dialkyl peroxides 118, 119, 176
Diazoalkanes 174
Diazonium salts 199
Disposable respirators 36

Disposal
  of chemicals 216, 252
  of wastes 208
Distribution of chemicals 77
Drain disposal 217
Drinking 13
Drugs 79
Drum(s) 113
  storage 81
Dry ice,
  see carbon dioxide, solid
Drying ovens 65
Dusts 89, 115
Dyes 176
Eating 13
Electric shock 48, 63, 68
Electrical apparatus 63
  safety 18
Electronic instruments 73
Embrittlement, hydrogen 142
Emergencies 215, 231
Emergency
  aid procedures 14
  equipment 39
  procedures 21, 40, 78
  reporting 42
Empty cylinders 135
Environmental rooms 60
Epileptic seizure 46
Equipment maintenance 16
Esters 169
Ethers 167
Evacuation 20
  procedures 41
Everyday hazards 20
Excess chemicals storage 203
Exhaust systems 50
Exits 18
Explosion 40, 42, 119, 127, 142, 144
  hazards 99, 151
Explosive(s) 79, 117, 161, 172, 187, 188, 194, 198, 209
  chemicals 115, 205, 246
  combinations 247
  compounds 121, 172
  gases 114
  limits 112
  peroxides 167
  substances 67
Explosivity 111, 189
Eye protection 3, 15, 21
Eyewash fountain 8, 26, 36, 40, 98
Eyewash station
  see eyewash fountain
Face shields 15, 21, 22, 90, 122
Fainting 48
Fire(s) 40, 42, 142
  blankets 30
  extinguishers 8, 26, 27, 39
  extinguishing systems 30
  hazard(s) 8, 99, 151
  hoses 30
  safety equipment 27
  suppression 214

General index

First aid	21, 40, 45, 63, 78	construction	56
equipment	39	design	56
kits	101	performance	57
procedures	14	Hot-air guns	68
Fittings	141	Hot-plates	68, 69
Flammability	88, 111, 207	Housekeeping	15
hazards	17	Hydrazine	192
Flammable		Hydrogen embrittlement	143
gases	83, 114, 135	Hydrogenation	137
limits	111, 112	catalysts	189
liquids	79, 80, 85, 113	Hydroperoxides	118, 119
solvents	60	Ice chests	15
substances	111, 113	Identification of laboratory waste	206
vapours	56	Ignition	113
Flash point	111, 112	energy	151
Floors	16	requirements	111
Flowers of sulfur	115	temperature	112
Food	66, 85	Impact strength	143
handling	15	Implosion	144
Foot protection	26	Incinerate	198
Footwear	15, 24	Incineration	107, 168, 182, 211, 220, 232, 238, 239
Formaldehyde	170		
Fume		off-site	224
extraction systems	26	on-site	225
hoods	162	Incinerator(s)	157, 163, 177, 182, 209, 227
Fungicides	237		
Furnaces	68	Incompatible	
Gas		chemicals	129, 160, 205, 208, 248
cylinders	44, 82, 130, 196	gases	133
lines	132	materials	214
Gavage	108	substances	85
Glass		wastes	214, 235
equipment	139	Injection	89, 91, 108
vessels	144	Inorganic	
Glassware	17, 66	acid halides	241
Glove(s)	15, 17, 18, 22, 103, 122, 143, 152	acids	44
		bases	44
box	59, 102, 105, 107	cyanides	170, 186
materials	23	fluorides	186
Goggles	22, 147, 149	hydroperoxides	182
Guards	16, 122	peroxides	182, 207
Hair dryers	73	sulfides	185
Halides of non-metals	241	wastes	177
Halogenated		Insecticides	237
hydrocarbons	164, 188	Insoluble sulfides	180
wastes	209	Inspection	214
Handling procedures	93, 96	Inventory, of chemicals	85
Hazardous wastes	207	Isolation	
Health and hygiene	15	areas	41
Heat guns	73, 124	rooms	59
Heating		Lab pack(s)	208, 228, 232, 235
baths	67	Labelling	203, 207
devices	68	for hazardous chemical transport	235
mantles	68, 71	Labels	78
tapes	68	Laboratory	
Heavy metals	102, 177, 249	coats	25
chlorates(VII)	199	hoods	52
HEPA filters		managers	202
see High efficiency particulate air filters		personnel	202
		sinks	26
High efficiency particulate air filters	96, 105, 107, 108	supervisors	202
		Laboratory-owned cylinders	136
High voltage circuits	73	Lactams	244
Hood(s)	52, 84, 102, 107, 144	Landfills	157, 163, 165, 167,

Index (cont.)

Landfills (Contd)   168, 172, 182, 188, 190, 209, 211, 221, 227, 232
- sanitary   232
- secure   232

Laser
- hazards   22
- operations   18

Light metals   177
Limits of flammability   112
Liquefied gases   114, 142, 145
Liquid
- air   18, 65, 115
- gases   114, 134, 138, 145
- nitrogen   18, 65, 115, 127

Local exhaust systems   58
Low temperature equipment   143
Magnetic stirrers   68
Medical
- emergencies   40
- facilities   41

Mercury poisoning   106
Metal(s)   248
- acetylacetonates   192
- acetylides   126, 146
- alkanecarboxylates   192
- alkoxides   192, 244
- alkyls   29, 30, 241, 245
- amides   179, 182, 188, 244
- aryls   245
- azides   187
- carbonyls   192, 245
- hydrides   29, 79, 82, 184, 245, 248
- nitrates   184
- phthalocyanines   192
- powders   245

Methyl ketones   172
Mixing devices   67
Molybdates(VI)   182
Motors   64, 67
Mouth suction   15
Neck injuries   46
Needles and syringes   79
Neutralizing agents   43, 44
Nitriles   169
Nitro compounds   172
N-Nitrosamides   174
N-Nitroso compounds   173
Noble metals   159
Non-halogenated wastes   209
Non-hazardous laboratory wastes   233, 234
Non-metal
- alkyls   193, 245
- aryls   193
- hydrides   190, 245

Oil
- baths   68, 71
- sludges   221

Open flame   17
Organic   247
- acid halides   241
- acids   168
- acyl halides   249
- anhydrides   249
- compounds   248
- halogen compounds   249
- hydrocarbons   223
- hydroperoxides   184, 199
- nitro compounds   249
- peroxides   117, 174, 176, 199

Organization, safety and health   202
Organo-
- inorganic chemicals   192
- mercury compounds   193
- metallic compounds   161, 192, 218
- sulfur compounds   168

Orphan reaction mixtures   194, 203
Outlet receptacles   63
Overload protection   64
Oxidizable substances   149
Oxidizing
- agents   86, 92, 182, 196, 249
- gases   136

Packaging of waste   228
Peroxide(s)   117, 128, 159, 174, 194
- content   118
- forming chemicals   203, 205, 242
- reagents   175

Peroxidizable compounds   243
Personal air-sampling   58
Phenols   165
Pigments   176
Pilot-plants   113
Piping   141, 146
Plastic equipment   141
Poisoning   106
Polychlorinated biphenyls   164
Polycyclic aromatic hydrocarbons   104
Polyhalogenated compounds   225
Polymerization   99, 137, 186, 245
- catalysts   82

Polymers   218
Polynitro
- compounds   173
- aromatic compounds   199, 225

Precipitation of
- cations   252
- metal anions   252
- metal hydroxides   251
- metal oxides   251
- sulfides   183

Pregnancy   91
Pressure
- gauges   139
- regulators   134, 147
- relief devices   138
- vessels   136, 137

Procurement of chemicals   77
Protective
- apparel   4, 15, 21, 25, 78, 94, 122, 152
- clothing   231
- devices   122
- equipment   90, 98

Pulmonary resuscitation   45
Pumps   67
Pyrophoric
- agents   189, 191, 193

General index

chemicals 203, 205, 245
hydrides 184
Quaternary ammonium salts 238
Radioactive materials 79, 237
Radioactivity 177
  hazards 18
Re-use of laboratory chemicals 157
Reactive chemicals 79, 115
Reactivity 177, 207
Record keeping 215
Recovery of
  laboratory chemicals 157
  solvents 159
  valuable metals 157
Recycling of laboratory chemicals 157
Refrigerators 15, 66, 85
Regulating valve 134
Regulators 134
Relief
  devices 136
  valves 138
Respirators 31
Respiratory
  equipment 152
  hazards 16
  protection 231
  protective equipment 31, 149, 153
  tract 97
Rotary evaporators 67
Routes of exposure 89
Rubber gloves 23
Safe disposal 65
Safety
  awareness 13
  belt 39
  data sheets 77
  equipment 14, 21, 26, 78, 94
  general recommendations 13
  glasses 90, 147
  goggles 90, 152
  shields 38
  showers 18, 26, 36, 40, 98
  training 15
Sanitary
  landfill 232
  sewage systems 163, 208, 216
Sealed tube reactions 141
Secure landfill 232
Segregation of laboratory waste 206
Self-contained breathing
  apparatus 31, 39, 40, 153
Separating funnels 17
Sewage systems
  sanitary 163, 208, 216
Shakers 67
Shields 16, 122, 137
Shock 46
Shock-sensitive compounds 121, 246
Shutdown procedures 41
Skin absorption 78, 186
Smoke tubes 57
Smoking 13, 15
Solidification of liquid wastes 235
Solubility 217

Solvents 159
Space heaters 68
Spectacles 21
Spillage(s) 26, 78, 94, 95, 97, 98, 106, 107, 121, 158
  kits 43
Spilled
  chemicals 16
  liquids 44
  solids 44
Splash goggles 22
Spontaneous
  combustion 112
  ignition 112
Starch 234
Start-up procedures 41
Stirrers 67
Stockrooms 80
  personnel 202
Stoppers 67
Storage cabinets 52, 86
Storage of
  chemicals 77, 79
  laboratory waste 213
Storage trays 84
Stretchers 39
Strong acids 92
Strong bases 92
Substituted hydrazines 192
Sulfonyl halides 168
Supplied air respirators 31, 60
Surplus chemicals 201
Syringes 91
Systematic identification 194
Thermometers 66, 106
Thread lubricant 136
Threshold limit value 89
Toxic
  compounds 161
  gases 79, 152
  hazards 88
  substances 67, 82, 86
  vapours 52
Toxicity
  cations 178
  developmental 91
  testicular 92
  types of 87
Toxicology 88
Training 215
Transport
  of chemicals 83
  of hazardous chemicals 227
  procedures 227
Tritium 55
Tubing 141
Ultraviolet light 22, 127
Un-needed chemicals 160
Unattended operations 19
Unconsciousness 48
Unlabelled
  chemicals 194
  containers 16, 194
Unusual hazards 14

Vacuum
  apparatus 145
  distillations 65
  pumps 65, 124, 144
  work 144
Variable transformers 69, 73
Vented canopies 52
Ventilation 50, 114
  systems 62
Viable organisms 238
Volatile acids 60
Warning signs 18
Waste
  accumulation 208
  categories 228
  chemicals 180, 201
  burning of 225
  containers 18
  from life-science laboratories 236
  incinerators 221
  storage 214
  disposal procedures 18
  management 201
  organization 202
  volume of 202
Water-reactive
  chemicals 44, 82, 203, 205, 241
  metal halides 190
Water-soluble 218
Worker exposure 58
Working alone 19